Zoophysiology Volume 32

Zoophysiology

Volumes already published in the series:

Volume 1: *P.J. Bentley*
Endocrines and Osmoregulation (1971)

Volume 2: *L. Irving*
Arctic Life of Birds and Mammals Including Man (1972)

Volume 3: *A.E. Needham*
The Significance of Zoochromes (1974)

Volume 4/5: *A.C. Neville*
Biology of the Arthropod Cuticle (1975)

Volume 6: *Schmidt-Koenig*
Migration and Homing in Animals (1975)

Volume 7: *E. Curio*
The Ethology of Predation (1976)

Volume 8: *W. Leuthold*
African Ungulates (1977)

Volume 9: *E.B. Edney*
Water Balance in Land Arthropods (1977)

Volume 10: *H.-U. Thiele*
Carabid Beetles in
Their Environments (1977)

Volume 11: *M.H.A. Keenleyside*
Diversity and Adaptation in
Fish Behaviour (1979)

Volume 12: *E. Skadhauge*
Osmoregulation in Birds (1981)

Volume 13: *S. Nilsson*
Autonomic Nerve Function in the Vertebrates (1983)

Volume 14: *A.D. Hasler*
Olfactory Imprinting and
Homing in Salmon (1983)

Volume 15: *T. Mann*
Spermatophores (1984)

Volume 16: *P. Bouverot*
Adaption of Altitude-Hypoxia
in Vertebrates (1985)

Volume 17: *R.J.F. Smith*
The Control of Fish Migration (1985)

Volume 18: *E. Gwinner*
Circannual Rhythms (1986)

Volume 19: *J. C. Rüegg*
Calcium and Muscle Activation (1986)

Volume 20: *J.-R. Truchot*
Comparative Aspects of Extracellular Acid-Base Balance (1987)

Volume 21: *A. Epple* and *J.E. Brinn*
The Comparative Physiology of the Pancreatic Islets (1987)

Volume 22: *W.H. Dantzler*
Comparative Physiology of the Vertebrate Kidney (1988)

Volume 23: *G.L. Kooyman*
Diverse Divers (1989)

Volume 24: *S. S. Guraya*
Ovariant Follicles in Reptiles and Birds (1989)

Volume 25: *G.D. Pollak* and *J.H. Casseday*
The Neural Basis of Echolocation in Bats (1989)

Volume 26: *G.A. Manley*
Peripheral Hearing Mechanisms in Reptiles and Birds (1989)

Volume 27: *U.M. Norberg*
Vertebrate Flight (1990)

Volume 28: *M. Nikinmaa*
Vertebrate Red Blood Cells (1990)

Volume 29: *B. Kramer*
Electrocommunication in Teleost Fishes (1990)

Volume 30: *W. Peters*
Peritrophic Membranes (1991)

Volume 31: *M.S. Kaulenas*
Insert Accessory Reproductive Structures (1992)

Volume 32: *A.L. Val* and *V.M.F. de Almeida-Val*
Fishes of the Amazon and their Environment (1995)

A.L. Val V.M.F. de Almeida-Val

Fishes of the Amazon and Their Environment

Physiological and Biochemical Aspect

With 76 Figures

Springer

Dr. A.L. Val
Dr. V.M.F. de Almeida-Val
Instituto Nacional de Pesquisas da Amazônia
Alameda Cosme Ferreira, 1756
69083-000 Manaus, AM, Brazil

Cover drawings by Izeni Pires Farias

ISBN 3-540-58382-3 Springer-Verlag Berlin Heidelberg New York

Library of Congress Cataloging-in-Publication Data. Val, Adalberto Luís. Fishes of the Amazon and their environment: physiological and biochemical aspects/A.L. Val, V.M.F. de Almeida-Val. p. cm. – (Zoophysiology; v. 32) Includes bibliographical references (p.) and index. ISBN 3-540-58382-3 (Berlin: acid-free paper). – ISBN 0-387-58382-3 (New York: acid-free paper) 1. Fishes – Ecophysiology – Amazon River Watershed. 2. Fishes – Amazon River Watershed – Composition. 3. Fishes – Amazon River Watershed – Respiration. I. Almeida-Val, V.M.F. de (Vera M.F.), 1956– . II. Title. III. Series. QL639.1.V35 1995 597′.01′09811 – dc20 95-1001

Typesetting: Best-set Typesetter Ltd., Hong Kong

SPIN: 10426266 31/3130/SPS – 5 4 3 2 1 0 – Printed on acid-free paper

To Fernando and Pedro,
for the joy we have shared

Preface

The Amazon is a giant piece of "amphibian" land which is the result of complex geological and evolutionary processes. The number of living beings in such a land is difficult to estimate. The interactions between these organisms and the environment are fascinating but barely understood. These features lured us to the Amazon in 1981. However, soon after, we realized that the dimensions of these interactions were overwhelming.

This book is designed to review aspects of the physiology and biochemistry of fishes of the Amazon. The description of the pulsative nature of the environment and the distinct features of the ichthyofauna of the Amazon were central to the main goal. Nevertheless, any complete view is limited by the magnitude of the intraspecific variability coupled with the complex fluctuations of the environment. Thus, we have placed an emphasis on respiratory physiology and biochemistry. The reference list was made as complete as possible, particularly regarding special publications not readily available. We hope that this book is useful for comparative physiologists, tropical biologists, and the people interested in interactions between organisms and their environment.

We are grateful to many people who contributed to the making of this book. Our initial ideas were influenced by Drs. Arno Schwantes, Maria Luíza Schwantes, José Tundisi, Anna Emília Vazzoler, and Naercio Menezes. New insights, particularly with respect to the respiration as an integrated system, have been learned with Profs. Peter Hochachka, David Randall, and the late Grant Bartlett. Much of the information was derived from the reports of and the discussions with our students Elizabeth Araújo, Mércia Caraciolo, Oscar Costa, Izeni Farias, Sebastião Freire, Elizabeth Gusmão, Nádia Limeira, Jaydione Marcon, Lenise Mesquita, Plínio Monteiro, Márcia Moura, Nazaré Paula, Alberto Peixoto, Hernando Ramirez, Roberto Souza, and Cacilda Yano.

Prof. David Randall suggested we write this book during a meeting in Göttingen. Later, he made available to us all his facilities in Vancouver, always providing the necessary support, advice, and encouragement. As editor responsible for the book he made many valuable comments on the manuscript.

Finally, we are grateful to Patricia O'Byrne for her patient help in revising the manuscript, to Izeni Pires Farias and Jorge Soares Dácio for their remarkable line drawings, and to Nazaré Paula da Silva for her dedication in developing the photographs. The National Research Council of Brazil (CNPq), the National Institute of Amazon Research (INPA), and the Natural Science and Engineering Research Council of Canada (NSERC) have financially supported this work.

Manaus, Rainy season 1995

A.L. Val
V.M.F. de Almeida-Val

Contents

Chapter 1

Introduction

The world has been divided into the following faunal regions: (1) Ethiopian – most of Africa; (2) Oriental – tropical Asia and closely associated islands; (3) Paleartic – Eurasia above the tropics and northernmost Africa; (4) Neartic – North America; (5) Australia – Australia, New Zealand, New Guinea, Tasmania and a few smaller islands; (6) Neotropical – South and Central America. The Neotropical Region, containing the greatest diversity of freshwater fish fauna, is subdivided into eight regions including the Amazonian region. Located at the heart of this richness is the Amazon basin.

The first description of South American fish, including Brazilian fish, was published in Holland in the year 1648 by George Marcgrave, who had accompanied Prince Nassau to Brazil. Some of his descriptions were used by Linnaeus in the work, "Systema Naturae". By the 18th century, many Amazonian fishes were described by P. Artedi, M.E. Bloch, J.P. Bonaterre, L.T. Gronovius, A. von Humboldt, and B.G. Lacépède. However, it was the Brazilian, Alexandre Rodrigues Ferreira, working for the Portugese government, who made the major Amazonian fish collection. His manuscripts and the collected material were transferred from the Real Museum of Lisboa to the Museum National d'Histoire Naturelle in Paris by the French army supervised by Napoleon. This material never returned to Portugal or to Brasil and many descriptions presented by Georges du Cuvier in the next century were based on the material collected by A.R. Ferreira (reviewed by Rapp-Py-Daniel and Leão 1991).

The Amazon basin was visited by many naturalists including Johan Baptist von Spix, Carl Friederich von Martius, Johan Natterer, Robert Herman Schomburgk, and Francis de Castelnau during the beginning of the 1800s. During the second part of this century Louis Agassiz also visited the Amazon river and some tributaries with the Tayer Expedition (Böhlke et al. 1978). The beginning of the 1900s was characterized by many more foreign expeditions to the Amazon basin, including those directed by Allen, Eigenmann, Haseman, Schultz, and Ternetz (reviewed by Rapp-Py-Daniel and Leão 1991). The collected fishes were then distributed worldwide to many museums, some of them being destroyed during the second World War. Using the preserved material collected during those expeditions, many other naturalists have described fish from the Amazon.

During the 1900s, Alípio de Miranda Ribeiro, a Brazilian ichthyologist, made many expeditions to the interior of Brazil using the facilities provided by General Rondon. In contrast to the previous expeditions, the fishes collected by Ribeiro were deposited at the National Museum of Rio de Janeiro. After him, many other Brazilian naturalists and ichthyologists mounted expeditions throughout the country, including many places in the Amazon, which were previously inaccessible. The material collected so far has been deposited in the Zoology Museum of São Paulo University and, more recently, in the Fish Collection of the National Institute of Amazon Research (INPA).

Until 1960, nearly all studies focusing on Amazon fishes were conducted from systematic and/or taxonomic points of view; a few explored the biological and ecological characteristics. According to Lowe-McConnel (1987), it was between 1960 and 1970 that studies on the ecology of fishes of the Amazon started. Most of these studies were conducted outside the Amazon on a few animals collected from restricted areas. These studies, therefore, do not allow for many general conclusions on distribution, behavior and biogeography of fish to be made.

The extent of the Amazon region, the nature of the rivers, lakes, floodplain areas, water level changes, rain storms, along with the absence of facilities suitable for the support of scientific expeditions throughout the region, have made the studies on fishes of the Amazon very arduous. Nowadays, despite the existence of some scientific institutions in the Amazon, our knowledge is limited to results from repetitious studies applying specific methodologies to different fish species; in areas such as reproduction, growth, migration, feeding, and cultivation of Amazonian fishes.

The purpose of this book is to review the accessible information about Amazonian fishes and their environment. Thus, the second chapter will describe the general characteristics of Amazonian waters, emphasizing the environmental diversity from an abiotic point of view. Geological aspects of basin formation will be presented to support discussion of climate, hydrology and water types. These aspects have a direct influence on fish biology. The environmental characteristics that enabled the colonization of the Amazon basin by a few fish species, their existence, and evolution will be discussed according to recent theories. Daily and annual variations of many physical and chemical parameters of water will be reviewed in order to discuss their effects on fish physiology and biochemistry. Recent environmental modifications produced by human activities will be analyzed in order to facilitate predictions about their effect on Amazonian fishes.

Information about fish biology and diversity in the Amazon basin will be presented in Chapter 3. This chapter will also discuss several theories about the evolution of the extensive species diversity. Most of these theories accept an adaptational basis, i.e., most of the authors explain species diversity as the result of great environmental heterogeneity. By organizing the available information describing species biology and distribution, it will be possible to illustrate many adaptive strategies in feeding behaviour, migratory cycles and reproduction of the most representative species. Some new insights into biochemical variability (isozymes/allozymes) and chromosomal bands have enhanced the information about the evolution of Amazonian fishes. Chapter 3 will bring this information into a concise form in order to review our current knowledge of the relationships between these fishes and their environment.

Amazonian fishes have developed many ways to increase their capacity to survive in poorly oxygenated waters. Some species are virtually independent from dissolved oxygen because they are able to breathe air. Other species, which are normally water breathers, may use a variety of facultative adaptations for the uptake of oxygen. When concentrations of dissolved oxygen drop to critical levels, the strict water breathers show many physiological and biochemical modifications; some can survive under anoxic conditions. Chapter 4 will review the available data concerning anatomical, morphological, physiological, and biochemical aspects of respiration of Amazonian fishes. Thus, we will discuss the three fundamental respiratory types from an evolutionary point of view, emphasizing the effects of daily and annual oxygen variations on hematological parameters, intraerythrocytic phosphates, hemoglobin fractions, and blood-oxygen affinity. Some effects of high water hydrogen sulphide concentration, oil contamination and low water pH on fish respiration will be considered.

The physiology and biochemistry of Amazonian fishes were practically unknown before the Alpha-Helix expeditions (Hochachka and Randall 1978; Riggs 1979; Val 1986). After these expeditions, there was an increase in interest because the resulting publications revealed several fascinating new features of aquatic and aerial respiration. The Alpha-Helix expeditions had, as a first goal, the investigation of physiological and biochemical characteristics of the aquatic-aerial breathing transition. However, the reports suggested that lack of oxygen was not the only problem for Amazonian aquatic animals and that air-breathing was not the main evolutionary solution to the problem. They concluded that water temperature and pH represent a big environmental pressure for Amazonian fish biology. Chapter 5 will review the available data on adaptive biochemical features of Amazon fishes facing low levels of dissolved oxygen and will review aspects of the respiratory metabolism of air- and water-breathers. The effects of oxygen concentration, temperature, and pH on metabolic characteristics of Amazonian fishes will be emphasized. Finally, in the last chapter we will review the main information presented in the previous chapters, suggesting new insights for studies and projects which will lead to a deeper understanding of the relationship between fish and their environment.

Chapter 2

Amazonian Waters

The scenic grandeur of the Amazon riverways have impressed everyone including even the south Brazilian visitors. From some tourists observing the Rio Negro it has been heard that "this river looks like a sea of coffee". This expression emphasizes the turbidity of the water and the extensive size of a single tributary of the Rio Amazonas. The European first to have sailed in the Amazon river is considered to be Vicente Yañez Pinzon, ex-commander of the caravel Niña during Columbus' historical first voyage, who sailed up the river for 50 miles by the year 1500 (Klammer 1984). However, it was Don Francisco de Orellana who discovered the Rio Amazonas (Fig. 2.1). Typical of the first explorers, Gonçalo Pizarro, with Orellana as a crew member, departed from Quito, located in the high Andes, in February 1541 to search for the land of "El Dorado", the legendary golden kingdom. Running out of food, Orellana and part of the crew went searching for supplies on land. Finally, by the end of 1541, no longer able to proceed on land, they built a boat at the confluence of the Rio Napo and the Rio Aguárico and followed the river current. After sailing the unknown his boat reached the mouth of the river on the Atlantic coast by August 26, 1542. He did not know that he was sailing the Rio Amazonas which was first called "Rio de Orellana" (reviewed by Sioli 1984a). According to Klammer (1984), the first scientist to travel the entire length of the Amazon from Pongo de Manseriche to Belém, Pará state (see Fig. 2.1) was Ch. M. de La Condamine who first documented astronomical determinations of longitude in surveying the river.

Our current knowledge of the physical, chemical, and biological characteristics of these huge water tracts and their interactions, has demonstrated that they are intimately correlated with the geological history of the basin. All of these characteristics have affected the evolution of the fauna of the Amazon, so that it is important to have a general view of the characteristics of the basin in order to understand not only the biological diversity per se but also the specific ways in which the organisms are able to cope with this complex environment.

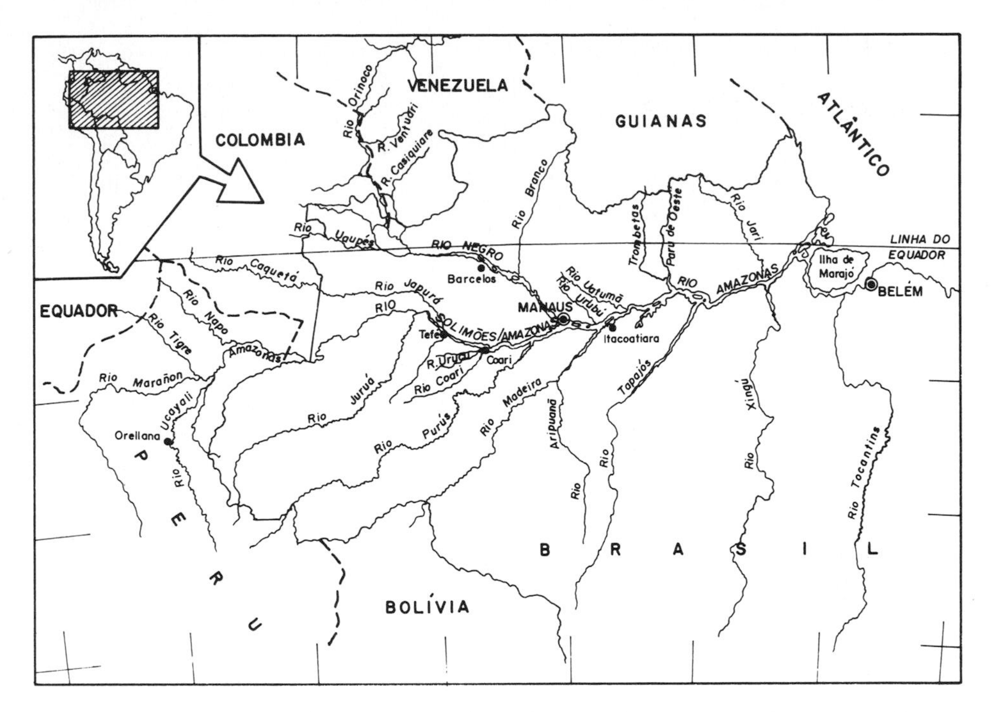

Fig. 2.1. The Amazon region. Note that the region extends across several countries. About 65% of the total extension of the Amazon is located within Brazil. The Amazon river has several denominations from its origin in Peru to its mouth in the Atlantic ocean

2.1 Geological Aspects of the Basin

The geology of the Amazon basin is known only in a broad sense (Bigarella and Ferreira 1985). The Amazon basin is an enormous valley framed by the Guiana Shield in the north and by the Central-Brazilian Shield in the south, both of which are part of the Precambrian Kraton (Putzer 1976). About $1.25 \times 10^6\,km^2$ of the area is occupied by sediments and the basin is elongated in shape. Oriented along an east-northeast axis, from the Peruvian and Colombian Andes to the mouth of the Amazon, the basin is about 3000 km long, 300 km wide in the eastern portion and 600–800 km in the western segment (Bigarella and Ferreira 1985).

The Amazon region has been submitted to the climatic changes of the last 2 billion years. The last consolidation of the land masses occurred 600 million years ago (Ma) (Putzer 1984). During the Paleozoic period, when the South American and the African continents were linked and when the other continents comprised what has been called "Old Pangea", sea water flowed through the region, probably reaching Peru and Bolivia. By the end of the Paleozoic, the Amazon basin, the Acre basin and the Maranhão basin were definitively incorporated into the continent (reviewed by Schubart 1983). Between 220 and 70 Ma, during the Mesozoic, a significant amount of sediment was accumulated and the rivers started to flow towards the east as does the Amazon river today, although the upper part of the Amazon (west from Manaus) probably drained to the west before the Andes uplift (Schubart 1983; Bigarella and Ferreira 1985).

The most extensive changes in the Amazon started by the end of the Mesozoic, Cretaceous period, climaxing in the Cenozoic. During this period the Gondwanaland breakup in the southern part of the hemisphere formed two continents, South America and Africa, and the Andes mountains were formed in the western part of the Amazon. Although Murphy and Nance (1992) surprisingly concluded that "every 500 million years the continents assemble into a single landmass", Bigarella and Ferreira (1985) refer to the Cenozoic as the "age of exceptional tectonic instability". In fact, the uplift of the Andes at the end of the Miocene induced stupendous changes in the Amazon basin. The Pacific drainage of the upper tributaries of the Amazon was cut and the whole system became Atlantic oriented. Thus, as pointed out by Putzer (1984) "a very large marsh and lake region with many rivers was born". The Pacific-oriented drainage is the accepted explanation for the existence of some Pacific relics in today's Amazon, including some natural formations. For example, some characteristic animals from the Pacific coast (dolphins, fish, and others) are found in eastern Peru and in the Madeira river, and the cataracts of the Madeira river near Porto Velho also exhibit Pacific origins (Putzer 1984).

Changes in ocean water levels during the Quaternary period, significantly influenced the Amazon scenery. Intensive erosion was the main feature during the low ocean water levels due to glacial periods, while sedimentation and formation of huge lakes took place during the high water levels (Putzer 1984; Sioli 1991). Thus, the modern Amazon sediment cover was formed during this period, probably under similar present-day drainage conditions (Schubart 1983). The thickness of this sediment cover extends to 300 m and is known as the Alter do Chão or Barreiras formation.

Increases in the seawater level then dammed the Amazon rivers. Because an area of about 1 million km^2 is below the 100-m contourline (Fig. 2.2), water level rises reached the upper part of the rivers. As a consequence, the large beds of sediment-rich rivers (the Amazon river, for example) were sedimented and constitute what today is known as *várzea* (floodplain), while the sediment-poor rivers became deeper. For example, the *várzea* of the Amazon river is up to 100 km wide while the Negro river has a 100 m depth at Manaus, 80 m below sea level (Junk 1983).

The tropical environments, including the Amazon region, are more stable than their Temperate Zone counterparts (Leigh and Wright 1990). This stability is a consequence of many geological factors operating in the region. As pointed out by Ab'Sáber (1979), orographic factors played important roles in differentiating the regional climates, particularly during the upper

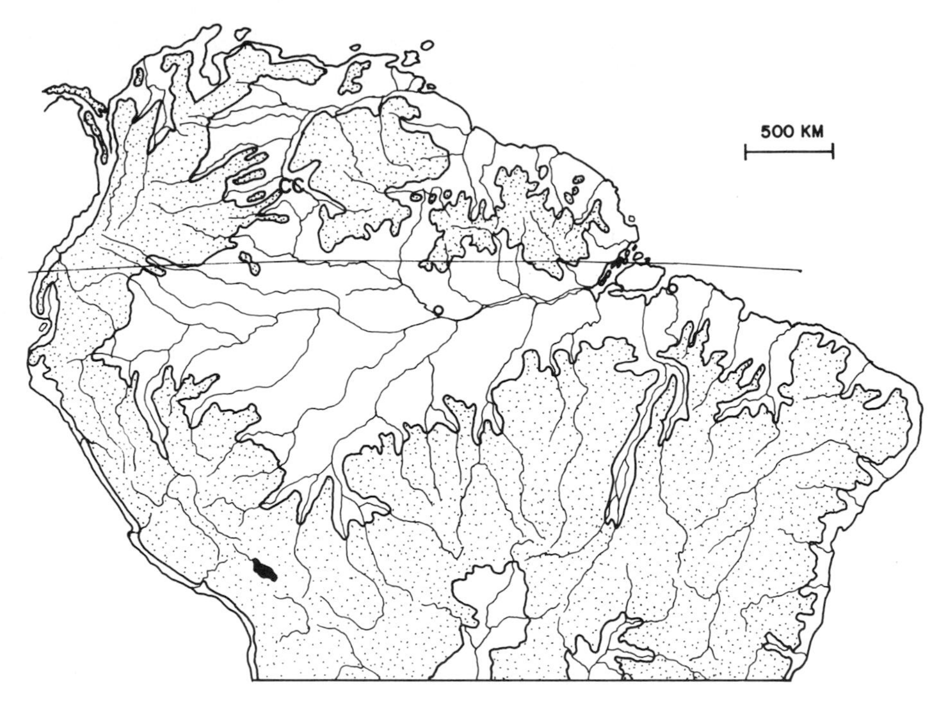

Fig. 2.2. A significant extent of the Amazon basin is below 100 m contourline (*open area*). Note that such a low altitude area reaches the Cassiquiare canal (*cc*), a natural connection between the Amazon and the Orinoco basin. (Redrawn from Schubart 1983)

Pleistocene, allowing the appearance of mosaics with different stable ecological characteristics. Thus, we will attempt to understand how the geological characteristics of the region, associated with its present-day environmental stability, have determined and maintained the current diversity of plants and animals in the Amazon.

2.2 Dimensions of the Region

The Amazon region has $7.5 \times 10^6\,km^2$, 65% ($4.8 \times 10^6\,km^2$) of which is located in the Brazilian territory. The rain forest covers about 80% of the Amazon region, i.e., $5.8 \times 10^6\,km^2$ (Salati and Vose 1984). The catchment area of the Amazon basin extends from 79° west (Chamaya river, Peru) to 46° west (Palma river, Brasil), and from 5° north (Cotingo river, Brasil) to 17° south (alto Araguaia, Brasil), according to Sioli (1984b). The Brazilian Amazon has 11 248 km in borderlines with seven countries (French Guiana, Suriname, Guiana, Venezuela, Colombia, Peru, and Bolivia). Thus, the Amazon system is the most tropical one that exists and, according to Sioli

(1984b), rivalled only by the Congo system in Africa ($3.69 \times 10^6 km^2$) which is only just over half the size of the Amazon.

2.3 Climate

The climatological characteristics of the Amazon rely on the geomorphology and geography of the region (reviewed by Salati 1985). Thus, the Central Plateau of the Amazon (4000 km in extension and a maximum slope of 100 m), which is bordered to the north by the Guiana shields, to the south by Brazil Central shields, and to the west by the Andean cordillera forming a "U" shape, receives the Atlantic hot and humid winds. These winds, interacting with other environmental factors including precipitation, solar radiation, water cycling, among others, explain, at least partially, the general climate pattern for the Amazon region. As pointed out by Walker (1991), evapotranspiration plays an important role since it represents up to 50% of the annual precipitation.

2.3.1 Climate in the Past and in the Future

It is reasonable to suppose that the climate in the Amazon has evolved as a consequence of the evolution of geological, biological, physical, and chemical characteristics. Analysis of the sediments associated with palynological studies of the Carajás mountains clearly suggest that the vegetation cover and consequently the climate, have changed considerably in that area (Absy et al. 1991). In addition, the existence of "stone lines" has been associated with dry periods during the upper Pleistocene (Ab'Sáber 1982). Recently, three main dry periods seem to have occurred in the Amazon, as suggested by C^{14} analysis: one at 4000, the other at 2100, and another at only 700 years B.P. (Anno Dommini 1200), as reported by Salati et al. (1991).

The future climate in the Amazon will depend on the world social and economic pressures. For example, if natural vegetation is converted into pastures, the temperature of the soil will increase by 3 °C, the precipitation will be reduced to about 640 mm/year (37% of the normal mean values) (Shukla et al. 1990), and many other environmental changes will occur (Buschbacher 1987; Buschbacher et al. 1987; Russel 1987). Increases in the atmospheric carbon dioxide concentration will also greatly effect climate changes (Woodwell et al. 1983). Thus, the climatological changes certainly had, still have, and will have significant effects on evolutionary processes of plants and animals of the Amazon.

2.3.2 Precipitation

The total annual precipitation varies from 1500 to over 3000 mm within the Amazon region (Fig. 2.3). Values above these ones are found close to the Atlantic coast (3250 mm) and in some places influenced by the Andes where values above 5000 mm occur (Salati 1985). Also, a significant seasonal variation in the maximum precipitation rates is observed. This pattern of precipitation is largely determined by the recirculation of the water vapour, i.e., part of the primary atmospheric water vapour precipitated as rain is retained in the forest and in the soil, evapotranspirated, and then returned to the atmosphere (Salati et al. 1991). About 54% of the total precipitation returns to the atmosphere as water vapour due to the effects of the forest and about 46% is drained through the rivers (Leopoldo et al. 1982; Salati 1985). This means that of the $11.87 \times 10^{12}\,m^3$ of rainwater that the Amazon receives per year, $6.43 \times 10^{12}\,m^3$ are evapotranspirated and $5.45 \times 10^{12}\,m^3$ run to Atlantic ocean through the Amazon river (Salati 1985). The type of vegetation cover, which is diverse in the region (Pires and Prance 1985), strongly influences the precipitation pattern in the Amazon.

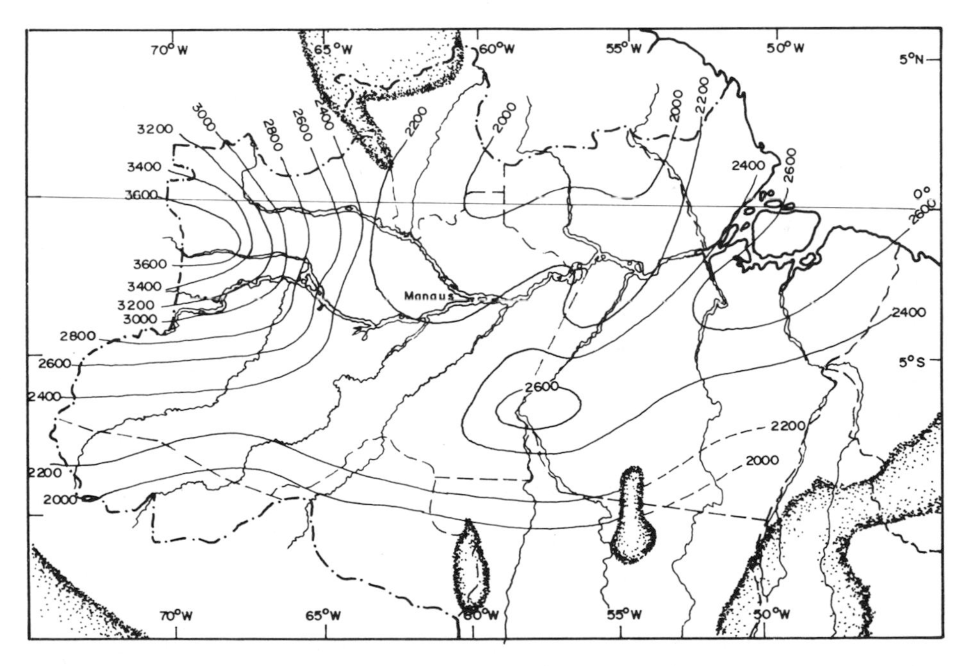

Fig. 2.3. Annual rainfall in the Amazon. Note that the total precipitation is irregular in the region. The isohyets indicate the average rainfall in millimeters. (Modified from Schubart 1983)

2.3.3 Solar Radiation

The solar energy that reaches the outer limit of the earth's atmosphere in Manaus city is practically constant throughout the year varying from 885 cal/cm^2/day in January to 767 cal/cm^2/day in June (Salati 1985). According to Ribeiro et al. (1982), about 50% of this radiation reaches the surface in Manaus and, although variable throughout the year, the cloud cover tends to be the main limiting factor.

2.3.4 Temperature

Although there is a high incidence of solar energy on the Amazon, the temperature is kept relatively low because of the existence of high levels of water vapour. According to Salati and Marques (1984), only about 30% of the available energy is used in local heating; the remainder being used in other physical and biological processes. Since the water vapour levels exhibit very low variations, the Amazon presents a tendency for isothermic conditions with monthly temperature means varying between 24 and 28°C. In addition, the daily temperature changes are larger than the annual ones.

A significant low temperature such as 14 °C is observed during a short period between May and June because of cool south polar winds blowing into the Central Amazon. This cooling phenomenon is known locally as *friagem* and has significant effects on ecological and limnological characteristics.

2.4 The Amazon Basin

The Amazon basin is comprised of innumerable rivers, lakes, *paranás* (channels), *igarapés* (small streams), beaches, *várzeas* (floodplain areas), and *igapós* (flooded forest). Nothing is permanent in the Amazon. The seasonal fluctuation in water level induces the appearance and disappearance of many formations including *paranás*, *igarapés*, *beaches*, and *igapós*; the *várzea* system shrinks and expands annually (Fig. 2.4). During the "dry" period, only the perennial riverine system persists. The common anastomosing characteristic of all these different ecosystems create a very low topographical relief producing a complex and extensive aquatic "landscape" which encompasses 7 million km^2 (Marlier 1967; Kramer et al. 1978; Sioli 1991). In addition, this immense water system pours 175 000 m^3 of water into the Atlantic ocean every second, representing about 20% of the total freshwater entering all oceans (Oltman 1967). This quantity represents 5 times that of the Congo and 12 times that of the Mississippi (Sioli 1984b). Interestingly, only one-third to one-half of this water volume, however,

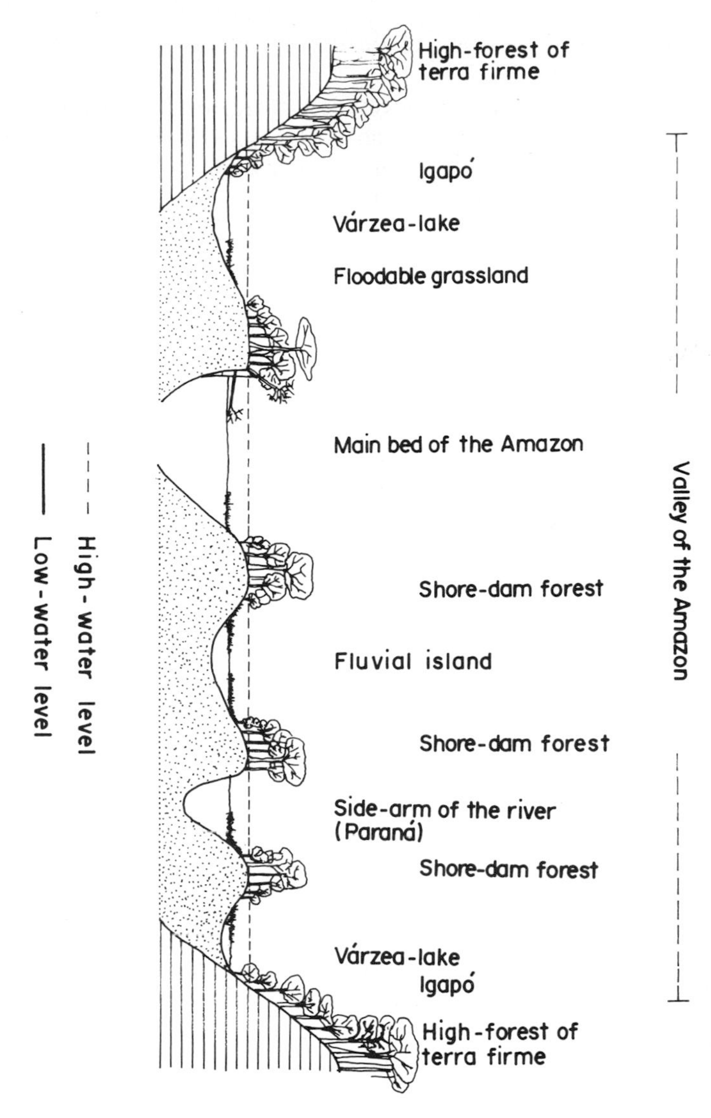

Fig. 2.4. The valley of the Amazon showing the main physiographic formations. The formations between the river banks (*terra firme*) are completely flooded during the high water level period. (Redrawn from Sioli 1984b)

originates from the high water precipitation in the Amazon region (Fink and Fink 1979; Sioli 1984b; Salati et al. 1991). As we shall see later in this chapter, this immense aquatic "landscape" is not homogeneous; its chemical, physical, and biological characteristics exhibit spatio-temporal variations.

2.4.1 Division of the Amazon Basin

From the foothills of the Andes to the mouth of the Amazon river in the Atlantic, it is possible to divide the basin into three main parts: the upper basin, the middle basin, and the estuary region (Sioli 1984b), each with very peculiar morphological characteristics of their rivers.

The upper basin is characterized by rivers with headwaters in the Andes. Due to the nature of the Andean environment, at times the rivers have clear water and at times turbid water, which is rich in suspended sediment. This is a consequence of landslides that often occur despite the existence of protecting vegetation. In the lowlands the rivers enter the forest and meandering begins; oxbow-lakes are formed. The soft sediments reach the Amazon river whose water is yellowish and turbid (Sioli 1984b). Thus, the *várzea* of white water rivers (see Sect. 2.4.3) of the upper basin is a kind of extension of the Andes (Fittkau 1975; Frailey et al. 1988).

Historical and economical reasons have resulted in a more complete characterization of the middle basin – historical mainly because of the colonization process (see Oliveira 1983) and economical due to developmental pressures (rubber and mineral exploration, hydroelectric power plants, and others). After having received almost all the Andean water the Solimões-Amazonas becomes an immense river which, in this part of the basin, does not meander. Except for the Negro river, many affluents pouring into the middle Solimões-Amazon have their headwaters in Guiana and Central Brazilian shields, geologically ancient zones. These affluents exhibit two divisions themselves: in the upper part, in the regions of hard rock soil, they flow "in 'normal', well-defined beds" often interspersed with rapids; in the lowland, entering the region of soft tertiary sediments, they become wider and rich in sediments (Sioli 1984b). Another important characteristic of the middle basin is the low river flow associated with the development of *várzeas*, particularly at the confluence of the Solimões-Amazonas, the Negro river, and the estuary.

As a consequence of the geological formation of the Central Amazon basin, waterfalls and rapids appear between the headwaters of the affluents in the Guiana and Central Brazilian shields, and the Solimões-Amazon river. These waterfalls and rapids represent an important physical barrier for fish dispersion in the Amazon basin (INPA/ENGE-RIO 1986).

The third part of the Amazon basin, the estuary region, starts near the confluence of the Xingú river. The enormous quantity of sediments carried into the Solimões-Amazon river is better observed at its mouth and on the Brazilian Guiana Atlantic coast, where an 80 km strip is a true, recent Amazon *várzea* (Sioli 1984b). A true delta is absent at the mouth of the Solimões-Amazon mainly because of the oceanic Brazilian current which does not allow the river to advance towards the open sea. However, many islands are observed at the mouth, the biggest one being the Marajó island with an area of almost 48 000 km^2. This island divides the mouth into two main parts: the north part, also filled with many smaller islands, is largely influenced by the ocean; the south part, formed by the Pará and the Tocantins rivers, has only a few small islands (Fig. 2.1). Tides are more important than the seasonal floods in controlling the water level in this part of the basin. Some rivers in this area exhibit the *pororoca* phenomenon (strong tidal waves). During this *pororoca* the riverbeds are disturbed by an invading strong tidal wave, producing a very strong distinct noise. The effect of seawater invasion is observed for as far as 1000 km from the mouth of the Solimões-Amazon river.

2.4.2 Basin Connections

Some peculiarities of the Amazon fauna, particularly fish fauna, have been attributed to physical connections of the Amazon river with other basins (Géry 1969, 1984). The Canal de Cassiquiare, for example, which connects the upper part of the Negro river (an affluent of the Amazon river) to the Orinoco river, Orinoco basin in Venezuela (Schubart 1983; Sioli 1984a), could have allowed exchanges between these basins. In addition, based on the distribution of characoid fish, Géry (1984) suggested that there were at least three past connections in Mato Grosso, Brazil, between the Amazon and La Plata basins.

2.4.3 Types of Water

Waters of varying characteristics can occur in the many different aquatic formations. Three general types of water occur in the Amazon: "white", "black", and "clear" (Sioli 1984b), although many intermediary patterns are observed. "White" waters characteristically flowing out of the Andes, are loaded with a large amount of suspended silt that confers a kind of pale muddy color to the water. The whole Solimões-Amazon river exhibits this type of water in spite of the other water types it receives from various tributaries.

Soil types, groundwater level, and vegetation are the main factors determining the differences between "black" and "clear" waters. They are present in the rivers draining from both the Brazilian and Guiana shields. Although they are distinct in their pronounced forms, they present many intergradations into each other. "Black" water is black in situ, but present a high transparency (1.3–2.9 m) compared to "white" waters (0.1–0.5 m); it is olive-brown or even a coffee-brown color in a glass. The black color is thought to be a consequence of humic and fulvic acids and other secondary metabolic compounds of plants leached from poorly drained soils which remain swampy many months each years (Marlier 1973). The best examples of "black" water are from the Negro river and many of its tributaries.

"Clear" waters, although also draining poor soils, present higher mineral content and less humic and fulvic compounds than the "black" waters (Stark 1970; Junk and Furch 1985). The Xingú and Tapajós rivers are the major "clear" water rivers in the Amazon basin. In addition, the small *igarapés* in the middle basin normally have "clear" water during the rainy season and a dark colour during the dry season (Junk and Furch 1985). These different types of water also present different physical, chemical, and biological characteristics.

2.4.4 Chemical and Physical Characteristics of Amazon Waters

Soluble inorganic and organic compounds are washed out of the vegetation by rainwater. These chemical substances, plus those washed out from the soil, will be drained into the water tracts. Thus, chemical, physical, and biological characteristics of water systems have been related to the properties of the vegetation cover and the soil that they drain. The vegetation cover and the soil in the Amazon basin are not homogeneous and, therefore, nor are the water systems.

Except by those bodies of water which originated in the Andes, such as the Solimões-Amazon river and the *várzea* lakes which can be classified as having carbonate waters, all other Amazonian systems are different from the "world average freshwater". However, according to Furch (1984), the Ca^{2+}/HCO_3^--poor waters exhibit a dominance of alkali metal and high trace element levels (Table 2.1). These characteristics, consistently observed for ion-poor water in the southern Amazon, are not paralleled by any other freshwater system in the world (Furch 1984).

Because of their different chemical composition the three major water types exhibit very characteristic water pHs (Table 2.1) and electrical conductivities. Higher pH values for "white" waters clearly suggest its higher buffer capacity when compared to "black" and "clear" waters. In addition, the conductivity observed for "black" waters is very low, $8\,\mu S/cm$ at 20 °C, compared to ca. $70\,\mu S/cm$ at 20 °C for "white" waters (Junk 1983). "Clear" waters have intermediary values for these parameters although they are very heterogeneous. It is important to emphasize that these parameters are strongly influenced by oscillations in water levels. Junk (1973), for example, showed an increase from 50 to $>300\,\mu S/cm$ in the total electrolytes in Xiborena lake due to sediment recycling.

Humic and fulvic acids also play important roles in defining the chemical and physical characteristics of water systems (Weninger 1985). For example, they can complex with metals and then with phosphorus, limiting this nutrient for phytoplankton (De Haan et al. 1990). Significant fluctuations in the levels of humic and fulvic acids are observed in the Amazon basin (Ertel et al. 1986).

Table 2.1. Mean ionic composition, specific conductivity (μS/cm), and pH in selected natural types of Amazon waters

	Solimões river[a]	Negro river[a]	Lago do Rei[b]	Forest streams[a]
Na (mg/l)	2.3 ± 0.8	0.380 ± 0.124	–	0.216 ± 0.058
K (mg/l)	0.9 ± 0.2	0.327 ± 0.107	–	0.150 ± 0.108
Mg (mg/l)	1.1 ± 0.2	0.114 ± 0.035	1.8	0.037 ± 0.015
Ca (mg/l)	7.2 ± 1.6	0.212 ± 0.066	8.3	0.038 ± 0.034
Cl (mg/l)	3.1 ± 2.1	1.7 ± 0.7	–	2.2 ± 0.4
Si (mg/l)	4.0 ± 0.9	2.0 ± 0.5	4.4	2.1 ± 0.5
Sr (μg/l)	37.8 ± 8.8	3.6 ± 1.0	–	1.4 ± 0.6
Ba (μg/l)	22.7 ± 5.9	8.1 ± 2.7	–	6.9 ± 2.9
Al (μg/l)	44 ± 37	112 ± 29	–	90 ± 36
Fe (μg/l)	109 ± 76	178 ± 58	–	98 ± 47
Mn (μg/l)	5.9 ± 5.1	9.0 ± 2.4	–	3.2 ± 1.2
Cu (μg/l)	2.4 ± 0.6	1.8 ± 0.5	–	1.5 ± 0.8
Zn (μg/l)	3.2 ± 1.5	4.1 ± 1.8	–	4.0 ± 3.3
Conductance	57 ± 8	9 ± 2	78.5	10 ± 3
pH	6.9 ± 0.4	5.1 ± 0.6	6.9	4.5 ± 0.2
Total P (μg/l)	105 ± 58	25 ± 17	900	10 + 7
Total C (mg/l)	13.5 ± 3.1	10.5 ± 1.3	–	8.7 ± 3.8
HCO_3-C (mg/l)	6.7 ± 0.8	1.7 ± 0.5	–	1.1 ± 0.4

[a] Data from Furch (1984).
[b] Data from Ribeiro and Darwich (1993).

Based on the primary productivity of phytoplankton and the standing stock of zooplankton, it is still impossible to make generalizations about the differences between "white" and "black" water bodies (Rai and Hill 1984a; Robertson and Hardy 1984). Although the effect of water chemistry on macrophyte growth is not fully understood, a clear effect of nutrient regime on the occurrence and growth rate of aquatic macrophytes is observed in the Amazon basin (Junk and Howard-Williams 1984). The same authors also suggest that acidic, nutrient-poor water, such as the "black" water systems, appear to restrict the occurrence of free-floating species to "white" systems. Recently, significant levels of *Chromobacterium violaceum* were reported to occur in "black" water systems. Such bacteria produce a pigment known as violacein which has strong antibiotic properties (Caldas 1990; Duran 1990). However, the in situ effects of such antibiotics remain unknown.

Water temperature in the Amazon basin is surprisingly constant, ranging from 29 ± 1 °C in the Solimões-Amazon river to 30 ± 1 °C in the Negro river year-round (Sioli 1984b, 1991), although large diurnal variations are commonly observed. During the *friagem* period water temperature in the lakes decreases significantly. Water density, as expected, is higher for "white" waters.

2.4.5 The Main Rivers

The enormous and extremely complex drainage system consists of an infinite number of water tract connected to a final collector, the Solimões-Amazon river, via the main tributaries. These aquatic ecosystems, carrying names which were coined by old Amazonians, include many different types of lakes, lagoons, cutoff river bends, *igarapés*, *paranás*, *furos* (small channels connecting water bodies), brooks, small rivers, *várzea* (floodplain area along the Solimões-Amazon and its tributaries), *igapó* (flooded forest), and *charcos* (marshy or swampy stretches) forming an extensive network whose total extensions exceed 1000 times that of the Solimões-Amazon itself. Thus, as pointed out by Sioli (1984b), this network implies that every single piece of land has very close contact with the water system.

The Amazon river entering Brazil is known as the Solimões river. It is, once again, called the Amazon river from the mouth of the Negro river to the ocean. The total length of the Amazon river, including the Ucayali river, is estimated to be 6518 km. Receiving so many tributaries, the Solimões-Amazon riverbed becomes wider and wider in order to accomodate the water it receives from the affluents as it progresses towards the Atlantic ocean. The river widens from 2 km near Iquitos to up to 6 km near Santarém, just after the confluence with Tapajós river, except at the Gorge of Óbidos, its narrowest place, when it has 1.8 km width and 100 m depth.

Despite the very gradual slope (about 0.001%) from its confluence with the Negro river to its mouth, the water current of the Solimões-Amazon is relatively strong, i.e., 0.5–1.0 m/s during the dry season and twice that value during high water levels which, according to Sioli (1984b), is a consequence of low levels of water friction in its large bed. This fact has enormous influence on the dispersion of fish eggs and alevins and on the river's sediment carrying capacity.

Among the main tributaries, the Negro river deserves special attention due to special characteristics which include the "black" waters, the presence of the Anavilhanas archipelago, the connection with the Orinoco river through the Cassiquiare canal, the lateral lakes and their *igapó*, and the absence of mosquitoes. Different from many other tributaries which have their headwaters out of the flat Amazon valley, the Negro river has its origin in a flooded region of about 100 000 km^2. The water moves slowly from this flooded area towards either the Orinoco basin or through the Negro river to the Solimões-Amazon river (Sioli 1984b).

Sedimentation zones are rare in the Negro river, however, after receiving the sediment-rich waters of the Branco river, an enormous sedimentation zone is observed in the lower part of the Negro river, which is known as Anavilhanas archipelago. The mouth of the river below this sedimentation zone behaves as lake. During the seasons when water levels are low, sandy clay material deposited in the bottom of the river emerges acting as small sedimentation zones. During the periods when the water level is high and the water current increases again, this material is carried away. The water level of the Negro river is directly affected by the fluctuations occuring in the Solimões-Amazon river, i.e., increases in water level of Solimões-Amazon river dams the Negro river in its own riverbed.

In contrast to the Amazon river, there is no real *várzea* along the Negro and other "black" water rivers. The floodplain areas, known as *igapó* (flooded forest), appear behind beautiful white sandy beaches submerged during the high water level season.

"Clear" water rivers follow the same general pattern of "black" water rivers. Tapajós river is the main example. Between its origin in the Central Brazilian shields and its mouth near Santarém city, the Tapajós river has many relatively small waterfalls. As with the Negro river, sandy banks are formed by the sediment carried in the bottom of the river. Leaving the Brazilian shields, the river enters the soft soil of the Amazon lowlands where its riverbed widens, the water current is slowed, and a sedimentation zone appears with many small elongated islands (Sioli 1991).

The Amazon river follows these three general patterns although many intermediary characteristics appear among them. However, it should be emphasized that because of an immense diversity and interaction of geological, physical, chemical, and biological conditions there are no identical environments in the Amazon.

2.4.6 *Várzea* and *Igapó*

Várzea is the floodplain area along the main "white" water rivers. It includes not only the main river channel but a complex of lakes, sidearms, small streams, and channels. These water tracts are strongly influenced by water level oscillation of the main river. During very high water levels it is even difficult to differentiate between these water systems. Behind the *várzea* appears a dense forest which is sometimes flooded as well, creating what is known as *igapó*. The *igapó* system is easily observed in "black" water systems because the forest reaches the river margins without a typical *várzea* formation between them.

The area over which the *várzea* normally occurs has been estimated to be 60 000 km^2 only in the Brazilian part of the Amazon basin (Martinelli 1986). The productivity of this area has been extensively evaluated (Ribeiro 1978; Welcomme 1979; Petrere 1983; Junk 1984b; Rai and Hill 1984a). From an ecological point of view the *várzea* represents an important source of food for fish alevins (Santos 1982; Ribeiro 1983). This is because the alevin and macrophyte productions are related (Bailey 1982).

Exhibiting very peculiar characteristics the *igapó* has received little attention compared to *várzea*. The *igapó* can be divided into three different zones: (1) the lower *igapó*, almost permanently flooded, has sandy quartz soil on the shores, is deficient of nutrients, and does not have a rich collection of species; (2) the middle *igapó*, regularly flooded, is characterized by complex trophic relations with a low-density, diversified fauna; and (3) the upper *igapó*, far from the shore, flooded only during seasons with exceptionally high water levels, represents a transition zone between the *igapó* per se and the terra firme (Adis 1984; Walker 1990). In addition, Walker (1990) suggested that the organisms inhabiting zones with irregular regimes (upper *igapó* zone) should exhibit physiological adaptations to the environmentally changing conditions, while those inhabiting zones with regular regimes (the lower and the middle *igapó* zones) might possess genetic adaptations.

2.4.7 Lakes

With a few exceptions, there are no true lakes (closed water bodies) in the Amazon. Among the exceptions are the Roraima lakes which are very shallow (2–3 m depth), egg-shaped, and filled up with rainwater during the rainy season. Their water is very low in nutrients and they almost completely dry out during the dry season (Junk 1983). These conditions produce very specific fauna and flora adapted to survive the seasonal fluctuations. These lakes, however, have received little attention from researchers.

Many other types of lakes are observed in the Amazon basin including oxbow lakes, terra firme lakes, lateral levee lakes, and *várzea* lakes. Oxbow lakes are often observed close to the margins of the meandering rivers. They are formed when the narrow end of a loop in a river is cut and closed off due to sedimentation. Old meanders are cut off and new ones are formed in a rapid process, which can be observed within a few decades. Nevertheless, this process may have many implications from a biological point of view, particularly regarding speciation events in organisms with a short life span, very few studies of these ecosystems have been developed (Sioli 1984b).

Ria lakes or terra firme lakes, as they are known in the Amazon, are formed due to sedimentation of the mouth by the main river, resulting in an enlargement of the lower portion of the river. This enlargement is sometimes very broad and deep as seen in "black" and "clear" water rivers because they did not fill their drowned lower courses with sediments. Affluents may discharge their waters into this enlarged part of the river sometimes through similar formations. The connection with the main river may disappear during the dry season, as occurs in the Cristalino lake. According to Junk (1984a), many of these so-called ria lakes exhibit lake-like conditions playing an important role in terms of energy conversion in the Amazon basin. They are, in general, highly influenced by the seasonal water level oscillations as are also the *várzea* lakes.

The excess of water poured into the Solimões-Amazon river overflows to the *várzea*, flooding the entire area up to the terra firme forest (see Fig. 2.5). The *várzea* lakes are filled during this process and return part of the

Fig. 2.5. **A** Lake Camaleão is filled during the high water level period. **B** The lake almost disappears during the low water level period

water stored during the flooding period to the main river where its own water level decreases. *Várzea* lakes are so far the best-studied aquatic formations in the Amazon although the studies have been restricted to the area close to Manaus city.

Chemical, physical, and biological studies on *várzea* lakes have shown that they have very high energy conversion rates, having high nutrient recycling rates, and increased life support capacity when compared to typical lotic environments (Schmidt 1973; Junk and Honda 1976; Junk 1979; Hardy 1980; Carvalho 1981; Bailey 1982; Furch et al. 1982; Furch 1984; Robertson and Hardy 1984). Thus, these lakes have important social and ecological implications (Junk 1979; Pereira 1991). For example, *várzea* lakes provide more than 90% of the total fish yield of Amazonian inland fisheries (Bailey 1981; Petrere 1983).

From an ecological point of view, *várzea* lakes are very interesting. The high productivity of these water bodies attracts many organisms which spend a long time feeding there. However, these organisms have developed many adaptations to cope with undesirable environmental conditions which, due to the high decomposition rate, include high levels of hydrogen sulphide and methane, and low oxygen availability, among others. In addition, floating macrophytes, having their roots immersed in a nutrient-rich water and their foliage above the water line, reproduce extensively and often cover the entire lake surface. This limits light diffusion, thus inducing a decrease in photosynthesis, and, as a consequence, an increase in the rate of the anaerobic processes. For some floating species, the biomass can increase up to 3000% per month (Junk 1984a). Associated with the immersed root of this floating vegetation is a rich fauna (Junk 1970; Sioli 1991).

As the water level starts to go down, the decomposition process is increased and undesirable conditions are intensified. Many organisms leave the local at this time, only some air-breathing fishes and those species which are highly adapted to hypoxic or even anoxic conditions remain. In addition, during the dry season, many of these bodies of water are greatly reduced or even disappear. Many fish species remain trapped inside these pools and while some estivate to avoid drying, others take advantage of the decreased competition for food, and others may even reproduce (Junk et al. 1983; Junk 1984a; pers. observ.). Then the river water levels start to rise and the whole cycle takes place again.

2.4.8 Annual Flood Cycle

The annual flood cycle is another important environmental factor that influences and shapes the Amazon basin. Virtually all living organisms in the Amazon are affected by a predictable annual flood cycle. The difference between the high and the low water levels range from 6 m (Iquitos, Peru), to 17 m (upper Japurá river), to 4 m (near the mouth). At the port of Manaus, an average crest of 10 m (Fig. 2.6), occasionally more, occurs on the main river channels (Schmidt 1973; Schubart 1983; Sioli 1984b).

The oscillation of river water levels is a direct consequence of the rainfall which is not regular in the Amazon (Fig. 2.3). The driest months in the middle basin are July, August, and September. The water current is very slow and the water leaving the Andes needs up to 3 months to arrive in the Amazon's estuary. The water, therefore, peaks at different times in the different parts to the final collector, the Solimões-Amazon river. Thus, the hydrological regime lags behind local climate conditions. In fact, at Manaus the rainfall typically ends in September and the river reaches its lowest level by the end of October (Junk 1970; Schmidt 1973; Sioli 1984b). In the small tributaries which are, however, strongly influenced by the local rainfall, many small peaks occur.

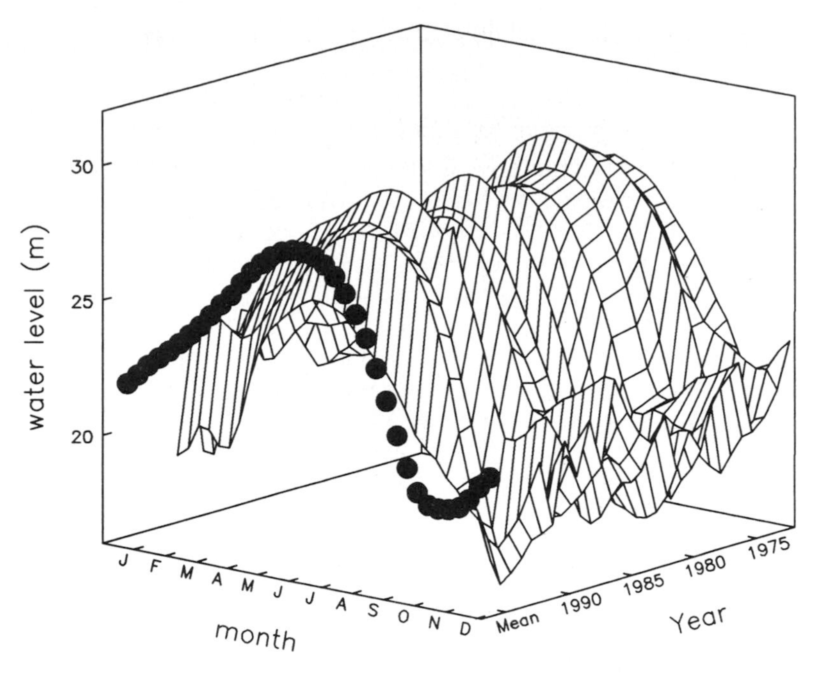

Fig. 2.6. Annual changes in water level at the port of Manaus obtained between 1972 and 1992. *Closed symbols* represent the 20-year average. Differences in intensity of water level changes are observed over different years. (Data from the "Capitania dos Portos-Manaus", AM)

Since the Solimões-Amazon valley is so flat, this huge annual flood inundates a significant portion of the vast floodplain area (*várzea*) of the lower and middle Amazon. As the water enters the *várzea*, it unloads its nutrient-rich sediment, significantly affecting the primary production of the lakes and the soil. This natural fertilization has many social, economical, and ecological implications (Junk 1980; Falesi 1986; Pereira 1991).

Water level oscillation, as the main environmental parameter for the Amazon, affects nearly all organic-aquatic environment interactions (Fink and Fink 1979; Almeida-Val 1986; Val 1986; Fernandes 1988; Junk et al. 1989; Walker 1990). Recently, Junk et al. (1989) proposed the flood pulse concept to characterize water level oscillations in river-floodplain systems, emphasizing that the flood pulse is the driving force of the biota in these systems.

2.4.9 Metabolism in Amazon Waters

Despite the significant amounts of organic carbon which are internally produced in "white" waters, heterotrophic processes dominate the metabolism in Amazon waters (Richey et al. 1990). Similar to other types of systems, the carbon produced by heterotrophic activity is derived from the *várzea* during the seasonal oscillation of river water level (Cuffney 1988). Part of this heterotrophic activity occurs anaerobically, producing significant levels of methane which is released into the river or into the atmosphere. This is due to typical O_2 limitations in the *várzea* (Tyler et al. 1987).

High heterotrophic activity is also observed in "black" and "clear" water streams. Walker (1990), studying immersed forest leafs in *igarapés* (the Tarumã-Mirim and Cuieiras rivers), showed that the nutrients generated in these systems are promptly assimilated by microorganisms (fungi, algae, and bacteria) and tree roots. This is reflected by the lower electrical conductivity in the water leaving these *igarapés*. She concluded that this process is very important to maintain the fauna of "black" and "clear" water *igarapés* and that there is no food available for filter feeders.

Physical, chemical, and biological processes, interacting in a complex way, determine the amount of dissolved oxygen in freshwater bodies. Such processes include photosynthesis and respiration, light diffusion, organic decomposition, and molecular O_2 diffusion, among others. These processes develop very peculiar particularities in different Amazon ecosystems because of the basin characteristics. For example, the water bodies on the Marchantaria island, located in the Solimões-Amazon about 15 km above its confluence with the Negro river, clearly illustrate this aspect. In fact, differences in shape, size, depth, winds, macrophyte cover, border vegetation, decomposition rates, and presence or absence of direct river communication, appear as determinants of dissolved O_2 concentrations in the different water bodies or even in different parts of the same water body on this island (Junk 1980; Junk et al. 1983).

Extreme variations in dissolved oxygen levels are very common in bodies of water in Temperate Zones. Although seasonal variation in dissolved O_2 is observed in tropical water systems, extreme variations tend to occur in a very short time. Oxygen levels can drop to zero at night in *várzea* lakes and reach high saturation levels at noon the very next day (Kramer et al. 1978; Junk et al. 1983; Val 1986). These rapid variations in O_2 availability have many biological implications, particularly on the respiratory patterns of aquatic animals.

During periods when the water level is high, when nutrients are unloaded into the *várzea*, development of an extensive floating macrophyte cover is observed, sometimes covering the entire body of water (Fig. 2.7). In these systems a weak O_2 cycle, whose levels are always below 50% saturation, is observed because of photosynthetic shading and heavy biological O_2 demand (Kramer et al. 1978; Junk 1984a). Similar situations are observed in the *igapó*.

Diurnal thermal stratification which is highest in the afternoon, can develop during high water level season resulting in hypoxic or even anoxic conditions in the lower water layers of *várzea* lakes (Junk 1984a). When the thermocline is broken down at the end of the day or by sudden winds, the

Fig. 2.7. The nutrients unloaded into the *várzea* lakes result in an extensive propagation of macrophytes which cover a significant extent of the water surface

water surface temperature decreases and the O_2 levels are reduced to as low as 10% of saturation. The small ponds and lakes which are unprotected by surrounding forest are more susceptible to periodic water turnover. The significant amount of hydrogen sulphide originating from decomposition activities in lower water layers, reaches the water surface during these turnover processes. Thus, in these cases, significant reduction of dissolved O_2 concentrations occur simultaneously with raised hydrogen sulphide levels. It is important to emphasize that different from temperate lakes, the terms epi-, meta-, and hypolimnion are inadequate to describe thermal stratification in tropical lakes (see Rai and Hill 1984b).

The small bodies of water, left by receding waters during the dry season, have a mixed pattern of dissolved O_2 extremes. Where larger quantities of macrophytes were accumulated, we see an intense decomposition process taking place and the biological demands for O_2 is so high, that even the water surface becomes anoxic. Simultaneously, a significant increase in hydrogen sulphide and methane is observed. Where such intense decomposition is not observed, the dissolved O_2 levels can reach more than 250% that of air saturation. This phenomenon is observed in some arms of Janauacá lake, although very hypoxic conditions do occur during unfavourable periods (Kramer et al. 1978). During these periods, significant extremes of temperature, such as 45 °C observed in some small pools during the dry period in the *várzea* of Marchantaria island (Val 1986), emphasize the effects of low O_2 availability on the aquatic organisms.

Horizontal microstratifications of oxygen concentrations near the water surface have been reported for some water formations in the *várzea* (Junk et al. 1983). The dissolved O_2 in the water surface, although in very low concentrations is, for many aquatic organisms, the sole source of O_2 during pronounced hypoxic or even anoxic conditions in the *várzea* (Table 2.2).

2.5 Recent Modifications Produced by the Activities of Man

The ecosystems of the Amazon are the result of a long-term environmental evolution in which chemical, physical, and biological features were and still are strongly influencing each other. The future of the Amazon region will be directly related to national and international economic interests in "developing" and exploring the region (Maués 1991). The endeavour of the Amazonian countries to enforce their own environmental protection laws, which are often well-conceived, will positively minimize the effects of these "developmental" projects. Undoubtedly, education will play an essential role as well.

Recent enterprises in the Amazon have induced many environmental modifications. These changes include pollution, use of natural resources at nonrenewable rates, movements of people, silting of water bodies, and damming of rivers, among others (Oliveira 1983; Sioli 1984c; Dourojeanni 1985; Treece 1989; Esteves et al. 1990; Fearnside 1989, 1990; Bittencourt 1991; Lacerda and Salomons 1991; Leite and Bittencourt 1991). Because the

Table 2.2. Changes in oxygen availability in some floodplain (*várzea*) habitats of the Amazon (Ilha da Marchantaria). Observe the variation of oxygen levels near the water surface (1–30 cm) and during different periods of the year at a given depth. I, V, and VI = First Lake in March, July, and November, respectively; II = Camaleão Lake, main channel in March; III and IV = Camaleão Lake in April, free water and among aquatic grasses, respectively

Depth (cm)	I[b]	II[b]	III[a]	IV[a]	V[b]	VI[b]
1	1.4	1.7	0.8	0.9	5.1	9.0
10	1.2	1.5	0.6	0.0	4.1	9.0
30	0.9	0.9	0.5	0.0	3.1	7.0
50	0.6	0.5	–	–	2.7	5.0
100	0.2	0.2	–	–	2.0	2.0
200	0.2	0.2	–	–	0.5	–

[a] Data compiled from Junk et al. (1983).
[b] Data compiled from Val (1986).

Amazon region is in essence, an amphibious landscape, any environmental modification on land immediately affects the aquatic ecosystems. It is very important to emphasize that even small changes can lead to real disasters because of the weak stability of the Amazonian ecosystems. However, most of these modifications have not yet reached significantly high levels.

The first and most common consequence of a "development" project in the Amazon is deforestation, although logged and managed areas still represent less than 5% of the total area of the Amazon region. During and following deforestation events, environmental deterioration including the appearance of many tropical diseases, silting of water bodies, climate and community structure changes is observed (Saldarriaga 1987; Fearnside 1989; Esteves et al. 1990; Noda and Noda 1990; Salati et al. 1991). In addition, CO_2 is released into the atmosphere when the forest is burned (Uhl and Buschbacher 1991; Victória et al. 1991) and nutrients are lost when trees are removed (Jordan 1987).

Although deforestation has a direct effect on fish biology, the hydroelectric power plants, the mercury contamination, and the petroleum exploration have also caused concern. The dams conceived for electricity generation in the Amazon, for example, have ignored the basic characteristics of the environment from both a biotic and an abiotic point of view. Balbina power plant, located at the Uatumã river, illustrates this issue. The Balbina dam flooded an area covering 2360 km^2 at a depth of 50 m (Brasil-Eletronorte 1987). Understandably, this drastic transformation of the landscape represents a significant environmental upheaval. In addition, increased levels of methane and hydrogen sulphide have been observed in the lake due to the decomposition of the flooded vegetation (Fearnside 1990). Simultaneously, the water leaving the turbines of Balbina hydroelectric power plant is deoxygenated mainly due to the nature of the water catchment system. Thus, fishes living upstream and downstream of the Balbina dam were affected. Many fish species have disappeared and others have had their habitat reduced in size. Considering that 80 more hydroelectric power plants have been planned for the Amazon basin (Fenraside 1990), it is important to intensify the stdudies of their effects on aquatic organisms, in particular, on fish populations.

Mercury contamination in the Amazon basin has increased significantly during the last decade mainly due to unregulated gold mining activities in the area. Brazil does not produce mercury; the metal is imported from European countries, reaching 340 tons in 1989 (Lacerda and Salomons 1991). According to Ferreira and Apel (1990), 50% of the mercury imported in 1989 was lost to the environment through gold mining activities. So, not surprisingly, many fish species contain significantly increased Hg levels in their tissues. The effects of mercury on physiological and biochemical processes in fishes of the Amazon are unknown. However, although changes in the environmental characteristics of the Amazon may result in slightly different effects, one may suppose that the Hg accumulation in fish of the

Amazon has the same general effects as in other freshwater fish. These effects include decreased plasma osmolality (Lock et al. 1981), decreased activity of lactate, pyruvate dehydrogenasese in muscle, succinic dehydrogenase in gill (Sastry and Rao 1981), and neurological depression (Hara et al. 1976), among others.

The crude oil obtained from the margins of the Urucú river, a southern tributary of the upper Amazon river (see Fig. 2.1), has been transported by small oil tankers which travel through the Tefé and Amazon rivers to Manaus to be processed. Crude oil spills and their effects on aquatic animals have been extensively reported in marine and estuarine environments (Buikema and Cairns 1984). However, the extent to which it affects tropical freshwater organisms is unknown. As the oil remains on the top of the water column it limits the sunlight filtering into the water. This reduces photosynthesis and consequently reduces the amount of dissolved oxygen. Preliminary, experiments comparing the effects of crude oil on gill and facultative air-breathing fishes from the Amazon showed high plasma lactate levels in both species. This suggests that even the air-breathing fish cannot take up O_2 in water containing a surface film of oil (Costa 1991).

2.5.1 Fish Culture in the Amazon

Fish culture in the Amazon has experienced many problems. First, the region has no tradition in fish culture and second, it is still widely considered that the idea of cultivating fish in the Amazon is uneconomic (Goulding 1979). Junk (1984b) presented interesting arguments against this idea.

We believe that there are enough economical and ecological reasons to establish fish cultures in the Amazon. For example, the increase in the number of registered fishing boats operating in the Amazon has not been proportional to the total increase in yield. However, in order to maintain the catch/boat constant, the fishing boats have resorted to making longer excursions. In addition, the availability of fish is not steady throughout the year, i.e., during the dry season there is a shortage of fish for marketing (Junk and Honda 1976; Petrere 1978a,b; Silva 1988). It is important to also mention that regardless of the great diversity of fish, only 13 species have a significant yield (Honda et al. 1975). According to Pereira et al. (1991), 18 species are responsible for 90% of the 300 000 tons of fish marketed in the region between 1979 and 1987. Four of these 18 species accounted for more than 60% of the total marketed. Fish culture can maintain the availability of fish in the market and would protect wild fish stocks. Thus, these aspects, plus the significant increment of the human population in the Amazon during the last two decades, undoubtedly justify fish culture in the region.

In addition to the above incentives, fish culture is feasible in the Amazon. Many experiments have shown that it is possible to control spawning by successive injections of pituitary extracts into several species (Graef, pers.

comm.) such as *Colossoma macropomum*, *Piaractus brachypomum*, *Brycon* cf. *cephalus*, and *Prochilodus* cf. *nigricans*. Optimum conditions for larval development, management and production of food have been achieved as well (Pereira et al. 1991). Based on growth rates, there are many species of fish, including herbivores, omnivores, and carnivores, which show great promise. Salvo-Souza and Val (1990), for example, raised *Arapaima gigas*, a carnivorous species, which achieved a weight of about 4000 g in 1 year. The young (between 12 and 50 g) were fed on freshwater shrimps, while those above 50 g were fed on pieces of fish with no commercial value. Experiments using pellets of common natural products (aquatic macrophytes, powder of many different fruits and seeds) to feed fish have provided promising results as well. For example, Saint-Paul et al. (1981) observed that up to 20% of the regular diet of *Brycon* could be changed to *Eichornia crassipes*' powder without any significant effect on growth rate.

Chapter 3

The Amazon Ichthyofauna

The number of living species in the world is estimated to be at least 4 or 5 million and could be as high as 10 or even 30 million as suggested by Fittkau (1985). According to Raven and Johnson (1986), there are 45000 living vertebrate species, a number which represents only a small portion of all living organisms. Estimates for the number of teleosts, the most successful group of vertebrates, range from ca. 20000 (Nelson 1984; Lauder and Liem 1983a) to 30000 (Starck 1978). Of the living species, 58% are marine, 41% live in freshwater, and 1% migrate between both habitats (Moyle and Cech 1982). Most of this ichthyofauna, the ostariophysans, originated in South America after the breakup of Gondwanaland but prior to the breakup of South America and Africa and then evolved independently on both continents (Novacek and Marshall 1976). This hypothesis was recently reanalyzed by Brooks and McLennan (1991), using concepts of cladistics. The authors produced two equally parsimonious area cladograms, having a consistency index of 81.8%. These cladograms strongly support Novacek and Marshall's general conclusion.

The fish fauna of the Amazon basin is far richer than that of any other river system. Although our current knowledge regarding systematic biology and phylogenetic relationships of fishes of the Amazon is relatively small, many authors have attempted to estimate the number of fish species living in this tropical basin. The suggested numbers range from 1300 to 2500 (Roberts 1972; Böhlke et al. 1978; Rapp-Py-Daniel and Leão 1991). However, these figures become underestimates as the number of sampled areas increase (Santos and Carvalho 1982; Vieira 1982; Santos et al. 1984; Lauzanne and Loubens 1985; Amadio 1986; Ferreira 1986; Ferreira et al. 1988; Goulding et al. 1988; Galvis et al. 1989; Santos 1991).

Jenkins (1976) pointed out that nearly 10% of the North American freshwater fish fauna remains undescribed regardless of the enormous effort expended. Undoubtedly, even more work than that required so far in North America will be necessary before we can give names in addition to providing information about their phylogeny and natural history, for all fish species in the Amazon. It is therefore impossible to even estimate the size of the gap in our knowledge of the identity and relationships of fish of the Amazon.

The current gap in our knowledge resulted from many factors including the immensity of the fish fauna; the extensive size of the basin; the number of different habitats; the geological and geographical characteristics of the basin; the explosive evolution in many groups of these fishes; the limitations of the current sampling methods; the environmental dynamics of the region; and the presence of many marine invaders (Kramer et al. 1978; Fink and Fink 1979; Junk 1984b; Almeida-Val et al. 1991b, 1993a; Rapp-Py-Daniel and Leão 1991; Santos 1991). Some of these factors will be considered later in this chapter.

3.1 Ichthyofauna Composition

The Amazon ichthyofauna has representatives of almost all groups of freshwater fishes (Table 3.1). These fish species colonized nearly all types of habitats in the immense basin, including the oxygen-depleted lakes (Junk et al. 1983; Monteiro et al. 1987; Affonso 1990; Val et al. 1992a) and the "black" water rivers, which are acidic (pH 3.8–4.9) and close to being distilled water (conductivity $20–30\,\mu S\,cm^{-1}$) (Lowe-McConnell 1987).

During the last two decades, studies of fish phylogenetic relationships have experienced an extreme revolution. These studies show that the evolutionary interrelationships of all living beings can be reflected in the interaction of many different fields including systematic biology, paleontology, genetics, biochemistry, ecology, and physiology (Whitt 1987; Berry and Jensen 1988; Stanley 1990; Ferguson and Allendorf 1991; Stock

Table 3.1. Comparative size of world and Amazon freshwater fish fauna. (Data compiled from Fowler 1951, Fink and Fink 1979; Géry 1984; Nelson 1984; Kullander 1986; Lowe-McConnell 1987)

Order	World			Amazon		
	Families	Genera	Species	Families	Genera	Species
Lepidosireniformes	2	2	5	1	1	1
Lamniformes	1	1	1	1	1	1
Rajiformes	3	9	70[a]	2	2	12
Osteoglossiformes	6	26	206	2	2	4
Clupeiformes	4	68	331	2	12	17
Characiformes	15	252	1335	12	229	1200
Siluriformes	31	400	2211	14	235	1000
Gymnotiformes	6	23	54	6	23	54
Batrachoidiformes	1	19	64	1	1	1
Cyprinodontiformes	13	120	845	5	13	30
Synbranchiformes	1	4	15	1	1	1
Perciformes	150	1367	7791	6	50	350
Pleuronectiformes	6	117	538	2	2	5
Tetraodontiformes	8	92	329	1	1	5(?)

[a] Not all freshwater.

1991). Unfortunately, very little is known about these characteristics in Amazonian fishes. Thus, unless explicit statements about relationships are made, the major groups are discussed in approximate order of phylogenetic origin, according to the available literature.

3.1.1 The Elasmobranchii

Sharks, rays, and skates represent the oldest group of fishes living in Amazon waters. They are cartilaginous fishes belonging to the class Chondrichthyes. One family and one species of shark (Carcharinidae, *Carcharinus leucas*) and at least 12 species belonging to the families Potamotrigonidae and Pristidae, which include rays, skates, and sawfish, occur in the Amazon basin (Thorson 1972; Fink and Fink 1979; Géry 1984). The genus *Potamotrygon* is the most common. They have a light yellow body with black spots and they tend to bury themselves in the soft mud sediments. Specimens with a disc diameter of 1 m have been collected (pers. observ.). The presence of these animals of marine origin living in the Amazon's freshwater could be thought of in at least two different ways. One is that the Elasmobranchii were already in the Amazon before the Andes uplift 70 Ma ago and became completely adapted to the new freshwater system; they have never left the region. Alternatively, one could think of the group as invaders of marine origin from long ago. The biology of the whole group is poorly known.

3.1.2 The Dipneusti

Lepidosiren paradoxa, locally known as pirambóia or pirarucúbóia, is on Amazonian species of Dipneusti, a group of true lung fishes. This group includes two other genera, i.e., the one Australian species, *Neoceratodus* and at least four African *Protopterus* sp. (*Protopterus aethiopicus*; *P. amphibius*; *P. annectens*; and *P. dolloi*). These animals are members of an ancient group of bony fishes, the Sarcopterygii or lobe-finned fishes, which were widespread and successful from the Devonian to the close of the Triassic (Fink and Fink 1979). As emphasized by Fink and Fink (1979) from a genealogical point of view, "this group of fish are more closely related to *Homo sapiens*, as descendants of sarcopterygian fishes, than to any other living fishes with the exception of *Latimeria*, the coelacanth". The pirambóia dig burrows and estivate during the low water periods (Carter and Beadle 1931a). The black body of this animal is very elongated, with two pairs of lobe fins, five gill arches, and four gill clefts. It can reach up to 1.5 m in length, moves slowly, and lives in areas where swampy conditions exist (Carter and Beadle 1931a, pers. observ.).

3.1.3 The Teleostei

This group of fishes is the most diversified and abundant when compared with all other vertebrates (Nelson 1984). The teleosts of the Amazon include 11 orders and about 2600 fish species. They are also by far the most diversified group of fishes living in the Amazon region. Characiformes and Siluriformes are the major orders with about 2200 fish species. However, living representatives of the orders Osteoglossiformes (ancient group) and Perciformes (specialized fishes) are exceptional for comparative studies because they experience the same environmental conditions as do many other fishes. According to Lowe-McConnell (1975), this diversity is the product of extensive adaptive radiation within these groups since the isolation of the continent during the Tertiary period.

3.1.3.1 Osteoglossiformes and Clupeiformes

Osteoglossiformes, or "bony-tongued" fish, are represented in the Amazon by three species. *Arapaima gigas*, locally known as pirarucú, is a species endemic to the Amazon basin. It is an obligatory air-breather and was formerly included in the Osteoglossidae along with *Osteoglossum bicirrhosum* and *O. ferreirai* who are water-breathers. Currently, *A. gigas* is classified in the Arapaimidae (Lauder and Liem 1983b). It is the largest known freshwater fish, reaching 3 m in length and 250 kg in weight and has a red coloration on ventral and caudal body regions (Lowe-McConnell 1975; Souza and Val 1990; Val et al. 1992b). *Osteoglossum* species, known locally as aruanã, are much smaller, reaching 1 m in length. These animals have elongated, laterally compressed bodies, which are covered entirely by iridescent sheen scales except for on the scaleless, bony head. The dorsal and anal fins of the aruanã are long and almost fuse with the caudal fin (Fig. 3.1). They are slow swimmers and are found most commonly in lake-like formations. The Osteoglossidae contains three other species as well, *Heterotis niloticus* (Africa), *Scleropages formosus*, and *S. leichardti* (northern Australia and adjacent areas).

The Clupeiformes are represented in the Amazon by 2 families: Clupeidae (herrings) with at least 5 species and Engraulidae (anchovies) with about 12 species. Although many of them are endemic to the Amazon basin, they are fish of marine origin (Lowe-McConnell 1975; Géry 1984). *Pellona castelnaena*, *P. flavipinnis*, *Ilisha amazonica*, *Engraulis*, *Pristigaster*, and *Anchoviella* are found most commonly in the major rivers of the Amazon. They have silvery, highly compressed bodies, and look very much like oversized sardines (Fink and Fink 1979; Géry 1984; Santos 1991). The biology of the Amazonian Clupeiformes if poorly known.

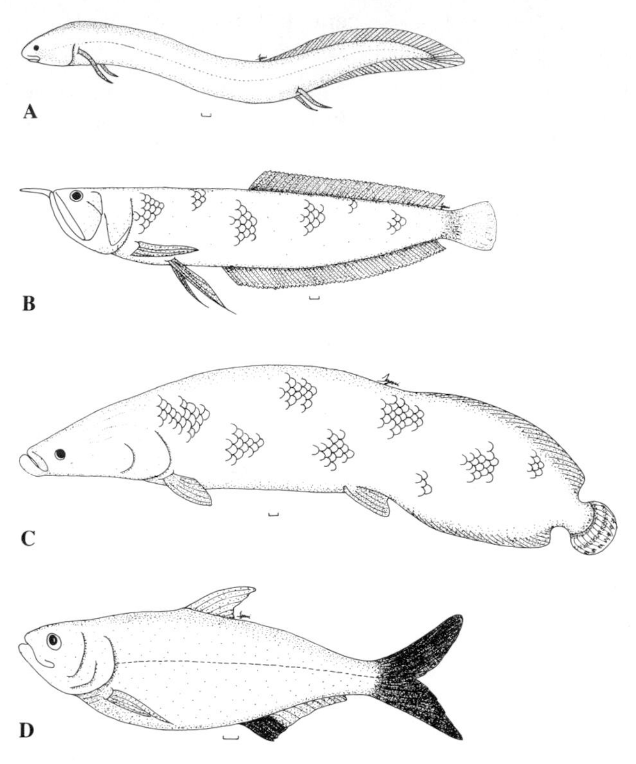

Fig. 3.1. Representatives of **A** Lepidosireniformes (pirambóia, *Lepidosiren paradoxa*); **B** Osteoglossiformes (aruanã, *Osteoglossum bicirrhosum*); **C** pirarucú, *Arapaima gigas*; and **D** Clupeiformes (apapá, *Pellona castelnaena*); *bar* = 1 cm

3.1.3.2 Ostariophysi

The Ostariophysi is the most dominant group of freshwater fishes in the world. The formation of the Weberian apparatus, a modification of four or more anterior vertebrae which connect the swim bladder to the inner ear for sound transmission, appears to be the unique characteristic which is shared by all members of this group. However, even though this modification provides the ostariophysans with greater hearing abilities, it is not well understood how the Weberian apparatus would have contributed to the success of the group. It is by far the largest teleost group of fishes living in the Amazon basin today. According to Géry (1984) all suborders of Ostariophysi are represented in the Amazon except for the Cypriniformes (carp-like fishes). The Amazonian ostariophysans include at least 12 families of Characiformes, 12 of Siluriformes and 6 of Gymnotiformes. According to Fink and Fink (1979) the South American ostariophysans have evolved from a few ancestors and have experienced a fantastic explosive radiation into almost all imaginable freshwater niches. Except for the marsupials of Australia, no other group of vertebrates has experienced such extensive specialization (Weitzman 1962).

Characiformes are considered to possess the most generalized gross body morphology of the Amazonian ostariophysans. However, they include animals with quite different shapes (Fig. 3.2) such as the freshwater dogfish (Raphiodontinae), the pencil fish (Lebiasinidae), and the piranhas (Serrasalmidae). The Amazonian Characiformes include about 1200 species distributed amongst 12 families (Table 3.2). Their distinctions are defined based mainly on trophic specializations and feeding structures (Fink and Fink 1979; Carvalho 1981; Santos 1982; Almeida 1984; Goulding 1985; Santos and Jegu 1989). This type of specialization, based on habitat partitioning, is poorly understood (Fink and Fink 1979). It is believed that this is the main reason that many more Characiformes will be discovered and described as other habitats are accessed, particularly the headwaters of the main rivers. Information on systematic biology, phylogenetic relationships, and ecology of some Characiformes of the Amazon has been presented by Böhlke et al. (1978); Fink and Fink (1979); Géry (1969, 1984); Santos (1980a,b); Hollanda (1982); and Nelson (1984) among others. Some specific characteristics of the main families of Amazonian Characiformes will be presented later in this book.

Siluriformes, the catfish, are the second most diverse and are among the most spectacular groups of fish found in the Amazon. This order includes 14 families and about 1000 fish species (Table 3.3). The size of these species ranges from the minute candirú (Trichomycteridae) of about 2 cm in length to the giant *Brachyplatystoma* (Pimelodidae) which reaches up to 3 m in length. They are scaleless or heavily covered by bony plates. Bony plates in distinctive patterns are found in some families such as Loricariidae, Callichthyidae, and Doradidae. Spines are often present on the anterior dorsal and pectoral fins, presumably for defense. Several species exhibit bottom-dwelling life habits and many are nocturnal. Many Trichomycteridae and Cetopsidae parasitize other fish species. Biological and ecological aspects of the siluriformes of the Amazon are poorly known; the available information has been reviewed recently by Zuanon (1990). Several species present interesting morphological adaptations to hypoxia as we shall see later.

The order Gymnotiformes, the electric eels, is the smallest group of Amazonian ostariophysans. It includes 6 families and about 54 species (Table 3.4), almost all of which are restricted to the Amazon basin. A few species (*Gymnotus carapo*, *Eigenmannia virescens*, *Sternopygus macrurus*, and at least four other species) are widely found in South America (see Mago-Leccia 1978; Géry 1984; Kramer 1990). Gymnotiformes and Siluriformes have many shared traits and they are now believed to represent a monophyletic lineage (see Fink and Fink 1981 for further details). Very little is known about the biology of this interesting group. *Electrophorus electricus* (Electrophoridae) reaches up to 2.5 m in length and is generally

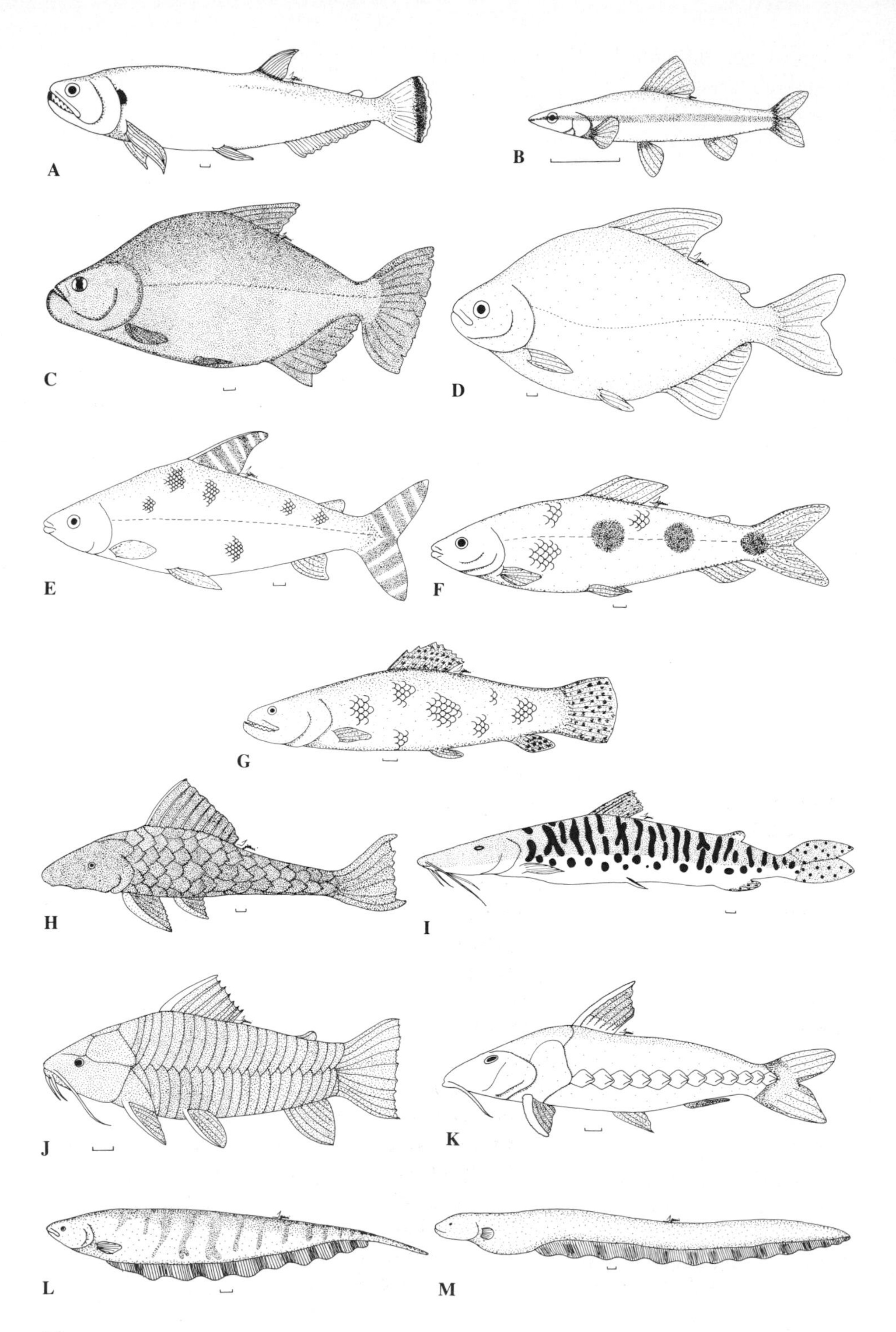
A
B
C
D
E
F
G
H
I
J
K
L
M

Table 3.2. Number of known species of Characiformes of the Amazon. The estimated number is based on the pertinent literature, which estimates the size of the order, and on the ratio of new/known species of new samples

Families	Number of species	
	Known	Estimated
Anostomidae	105	116
Characidae	776	882
Chilodontidae	3	3
Ctenoluciidae	4	4
Curimatidae	105	116
Cynodontidae	10	11
Erythrinidae	5	6
Gasteropelecidae	9	10
Hemiodontidae	50	55
Lebiasinidae	50	54
Prochilodontidae	30	32
Serrasalmidae	55	61

avoided by the natives of the Amazon due to its dangerous, powerful, electric shock. Aspects of the electrocommunication of Gymnotiformes have been recently reviewed by Kramer (1990). Aspects of systematic biology, geographic distribution, and phylogenetic relationships of some Gymnotiformes were studied by Schwassmann (1976); Mago-Leccia (1978); Lundberg and Stager (1985); Mago-Leccia et al. (1985); and Lundberg and Mago-Leccia (1986) among others.

3.1.3.3 Paracanthopterygii and Acanthopterygii

Paracantopterygii is represented in the Amazon basin by the minute *Thalassophryne* sp., family Batrachoididae, and order Batrachoidiformes. According to Géry (1984), representatives of this species were only recently

←

Fig. 3.2. A–G Representatives of Characiformes, **H–K** Siluriformes, and **L–M** Gymnotiformes of the Amazon. **A** Freshwater dogfish, peixe-cachorro (*Raphiodon vulpinus*, Cynodontidae); **B** pencil fish (*Nannostomus eques*, Lebiasinidae); **C** black piranha, piranha preta (*Serrasalmus rhombeus*, Serrasalmidae); **D** tambaqui (*Colossoma macropomum*, Serrasalmidae); **E** jaraqui (*Semaprochilodus insignis*, Prochilodontidae); **F** aracú (*Leporinus friderici*, Anostomidae); **G** traira (*Hoplias malabaricus*, Erythrinidae); **H** armoured catfish, acari-bodó (*Pterygoplichthys multiradiatus*, Loricariidae); **I** catfish, surubim (*Pseudoplatystoma fasciatum*, Pimelodidae); **J** tamoatá (*Hoplosternum littorale*, Callichthyidae); **K** cuiu-cuiu (*Oxydoras niger*, Doradidae); **L** electric eel, poraquê (*Electrophorus electricus*, Electrophoridae); and **M** electric eel, tuvira (*Gymnotus carapo*, Gymnotidae); *bar* = 1 cm

Table 3.3. Number of known species of Siluriformes of the Amazon. The estimated number is based on the pertinent literature, which estimates the size of the order, and on the ratio of new/known species of new samples

Families	Number of species	
	Known	Estimated
Auchenipteridae	60	64
Ageneiosidae	25	27
Aspredinidae	25	27
Astroblepidae	35	38
Callichthyidae	110	121
Cetopsidae	12	13
Doradidae	25	27
Helogenidae	4	4
Hypophthalmidae	1	1
Loricariidae	450	504
Pimelodiae	290	319
Trichomycteridae	175	192

collected near the Curuá-Una dam. Almost nothing is known about this group in the Amazon.

Acantopterygii is represented by many groups of fish including Cyprinodontiformes, Symbranchiformes, Pleuronectiformes, Tetraodontiformes, and Perciformes (Table 3.5). Many of them are extremely important in the Amazon from a commercial point of view. This group of fishes is the most recent in terms of phylogenetic origin and includes about 500 species (Fink and Fink 1979). Many Amazonian acantopterygians are fish of marine origin (Fink and Fink 1979; Géry 1984; Nelson 1984). The Perciformes, including the Scianidae, Nandidae, and Cichlidae, is a major group with

Table 3.4. Number of known species of Gymnotiformes of the Amazon. The estimated number is based on the pertinent literature, which estimates the size of the order, and on the ratio of new/known species of new samples

Families	Number of species	
	Known	Estimated
Apteronotidae	25	46
Electrophoridae	1	1
Hypopomidae	12	22
Gymnotidae	3	5
Ramphichthyidae	2	3
Sternopygidae	11	20

Table 3.5. Number of known species of Acanthopterygians of the Amazon. The estimated number is based on the pertinent literature, which estimates the size of the order, and on the ratio of new/known species of new samples

Famlies	Number of species	
	Known	Estimated
Batrachoidiformes	1	1
Cyprinodontiformes	30	34
Synbranchiformes	1	1
Pleuronectiformes	5	6
Tetraodontiformes	5	6
Perciformes	350	412

about 350 species, the Cichlidae being the most diverse family (Fig. 3.3). In fact, Cichlidae has the third largest number of species of any group of fishes in the Amazon (Géry 1984). This family includes the tucunaré (*Cichla ocellaris*) considered a delicacy in the Amazon, and the *Astronotus ocellatus* which has a very high resistance to hypoxia. Aspects of systematic biology, geographic distribution, and ecology of some Amazonian cichlids have been presented by Machado-Allison (1971); Fink and Fink (1979); Ferreira (1981); Lowe-McConnell (1987); Kullander and Ferreira (1988).

3.2 Feeding Behaviour

Food is essential in providing energy and nutrients for all biological activities of fishes, including the relationships among themselves and between them and their environment. Although many fish species persist for long periods of starvation and some do not eat during specific periods of their life cycle, food acquisition is necessary to build a metabolic reserve for such periods (Wootton 1979; Love 1980). Food acquisition by fishes is a process that generally involves searching, detection, capture, and ingestion (Keenleyside 1979). In addition, according to Wootton (1990), three basic questions must be answered during an ecological approach to the feeding characteristics of fish: what is eaten, when is it eaten, and how much is eaten?

Fish have been categorized into four main trophic categories: detritivores, scavengers, herbivores, and carnivores. Herbivores and carnivores are subdivided into many groups (Table 3.6) as proposed by Keenleyside (1979). Although this classification provides an important tool for the evaluation of feeding behaviour of fish, the study of these characteristics in fish living in an environment such as the Amazon basin have many other implications, most of them related to the pulsating nature of the environment. For

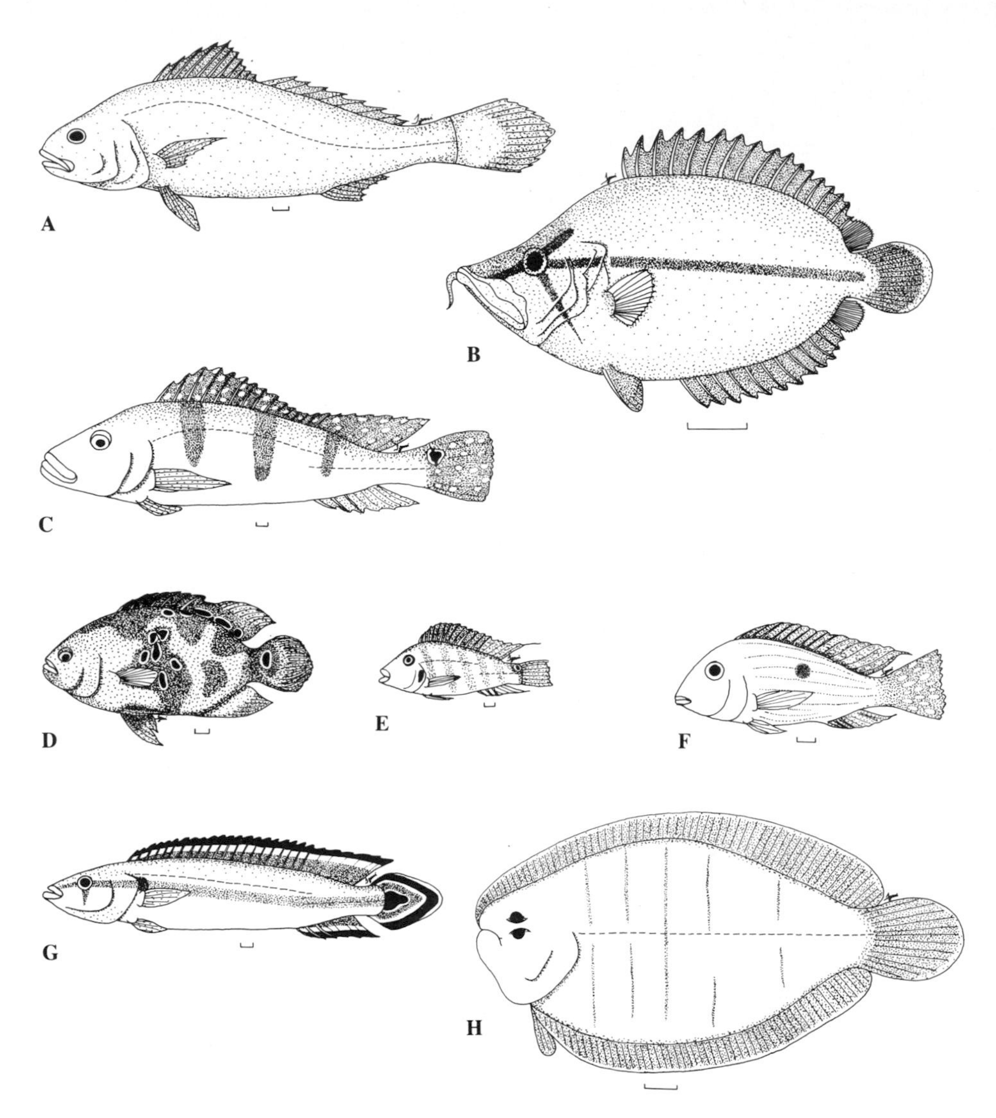

Fig. 3.3. A–G Representatives of Perciformes and **H** Pleuronectiformes of the Amazon. **A** Pescada-branca (*Plagioscion squamosissimus*, Scianidae); **B** leaf fish, peixe-folha (*Monocirrhus polyacanthus*, Nandidae); **C** tucunaré (*Cichla ocellaris*, Cichlidae); **D** acará-açu (*Astronotus ocellatus*, Cichlidae); **E** cará-bicudo (*Satanoperca jurupari*, Cichlidae); **F** acará-tinga (*Geophagus surinamensis*, Cichlidae); **G** jacundá (*Crenicichla* sp., Cichlidae); and **H** linguado (*Achirus* sp., Soleidae); *bar* = 1 cm

example, the food availability changes both in quality and quality as the water level oscillates (Goulding 1980; Junk, 1984b; Leite 1987; Lowe-McConnell 1987). Therefore, many fishes of the Amazon exhibit flexibility in their diets.

Table 3.6. Major trophic categories in fishes of the Amazon. The trophic categories of the mentioned species or group of species may change seasonally. See text for further explanations

Categories	Examples
1. Detritivores	*Curimata, Potamorhina, Psectrogaster, Curimatella Prochilodus, Semaprochilodus*
2. Herbivores	
2.1. Grazers	*Hypostomus* (= *Plecostomus*), *Pterygoplichthys*
2.2. Fruit eaters	*Colossoma, Piaractus, Myleus*
3. Carnivores	
3.1. Benthivores	*Serrasalmus*
3.2. Aerial feeders	*Osteoglossum*
3.3. Piscivores	*Serrasalmus, Pellona, Arapaima, Cichla, Acestrorhynchus*
3.4. Scale eaters	*Catoprion, Roeboides*
4. Planktivores	*Eigenmannina, Hypophthalmus, Chaetobranchopsis*

3.2.1 What Do the Amazonian Fish Eat?

The Amazonian fish feed on almost all available sources of food including invertebrates, aquatic macrophytes, spongi, bryozoans, fish, fruits, seeds, phytoplankton, detritus, and algae. Thus, all main trophic categories are represented among the Amazonian fish in spite of their opportunistic character (Marlier 1968; Ferreira 1981; Almeida 1984; Junk 1984b; Goulding 1985; Santos et al. 1991).

The flooded forest (*igapó*) represents an important source of food. During periods of high water levels, many fish species enter the forest where they eat fruits and seeds despite the low levels of dissolved O_2 available in these places (see Chap. 2). In addition, fruits and seeds form the foundation of the trophic chain in the "black" water systems (Santos et al. 1991). Among the fruit and seed eaters are many economically important species like those belonging to the genera *Colossoma*, *Mylossoma*, *Myleus*, *Brycon*, *Osteoglossum*, some members of the groups Anostomidae, Bryconinae, Cichlidae, Serrasalmidae (piranhas), and many catfishes of the families Doradidae, Auchenipteridae, and Pimelodidae (Aragão 1981; Junk 1984b; Goulding 1985; Araújo-Lima et al. 1986; Borges 1986; Leite 1987; Santos et al. 1991).

In general, fish feeding on similar types of food have similar traits either because they are phylogenetically related or due to convergent evolution of their morphologies (Wootton 1990). Although characins and catfishes of the Amazon are both seed and fruit eaters, they eat them in different ways;

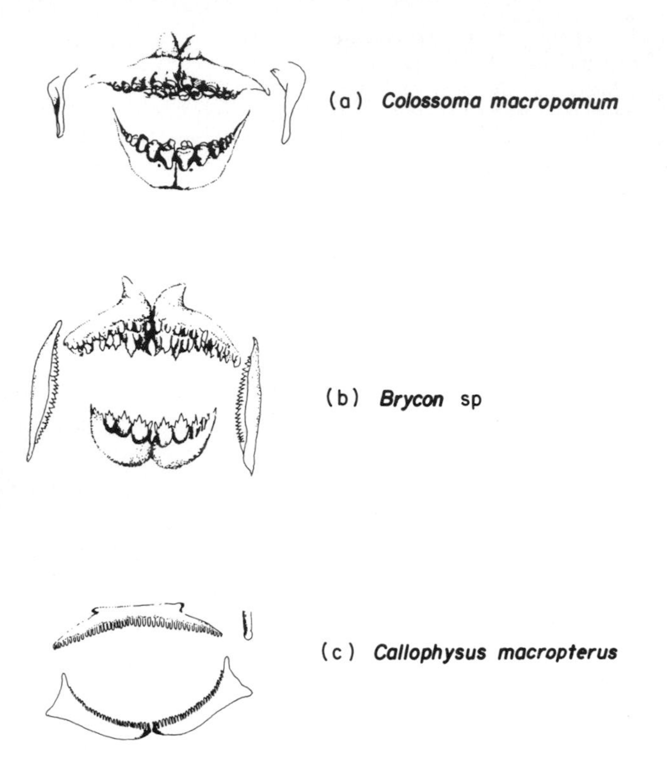

Fig. 3.4. Dentition of **a** *Colossoma macropomum*, **b** *Brycon* sp., and **c** *Callophysus macropterus*. (Redrawn from Goulding 1980)

characins with their well-developed dentition (*Colossoma macropomum*, for example, see Fig. 3.4) crush fruits and seeds, whereas catfishes normally grasp them (Goulding 1985). Many piranhas do not eat the whole seed; they appear to be only interested in endosperm material (Goulding 1985). The presence of fish species with quite different morphological traits feeding on similar source of food suggests a convergence of habits rather than of structures (teeth, jaw, intestine, enzymes, etc.). Why some fish species, having traits designed for other sources of food, feed on seeds and fruits is still unknown. However, the abundance of fruits and seeds in the flooded forest suggests that animals with this flexibility expend less energy in procuring food and are exposed to fewer feeding stresses.

Fishes of the Amazon also feed on other sources of vegetation including phytoplankton, roots, flowers, and leaves (including those of allochthonous origin), grass, algae, and aquatic macrophytes (Marlier 1968; Goulding 1980; Paixão 1980; Santos 1982; Almeida 1984; Junk 1984b; Leite 1987; Lowe-McConnell 1987; Zuanon 1990; Santos et al. 1991). However, except for a few species that are specialists (species of the genera *Ctenobrycon*, *Leporinus*, and *Poecilobrycon*, among others, Marlier 1968), these items are generally incorporated into the diet of herbivores, carnivores, detritivores, and scavengers.

Many fish species are piscivorous, feeding either on the whole prey, as do *Arapaima gigas* (Souza and Val 1990), or on small pieces bitten from the prey, as do some species of Serrasalmidae. Among the species which are strictly or mostly piscivorous, are some members of Characiformes, some cichlids, the clupeid *Pellona* sp., some scinanids, the electric eel *Electrophorus electricus*, and many catfishes (Marlier 1968; Knöppel 1970; Ferreira 1981; Junk 1984b; Leite 1987; Zuanon 1990). Some small characins (*Roeboides* sp.) and particularly one species of serrasalmid (piranha xidauá, *Catoprion mento*) are scala eaters (Goulding 1980; Santos 1991). Members of Cetopsidae and Trichomycteridae, the famous candirús, also feed on fish and/or parasitize them. In addition, both piranhas and candirús may cause significant damage to netted fishes (pers. observ.).

Different crustaceans, including Copepoda, Ostracoda, Cladocera, and palaemonid prawns, have been found in the stomach of many species of fish (Marlier 1968). Terrestrial or semi-aquatic arthropods constitute another important item for the diet of many fish. Fish often acquire insects from the flooded forest when the insects colonize floating aquatic macrophytes and/or when they fall from the canopy above. Ants and grasshoppers, for instance, are quite often found in the stomach of many species including *Osteoglossum bicirrhosum*, *Triportheus*, *Brycon*, and some cichlids (Marlier 1968; Goulding 1980; Santos et al. 1991).

A great amount of organic matter is often deposited on the floodplain areas. Consequently, it is not a surprise that several fish species of the Amazon feed on detritus, which includes a mosaic of food items. Among the Amazonian detritivores are many species of Curimatidae, Loricariidae, Callichthyidae, Prochilodontidae, and some Cichlidae (Leite 1987; Junk 1984b; Zuanon 1990; Santos et al. 1991). Araújo-Lima et al. (1986), analyzing the autotrophic source of carbon for detritivorous fishes of the Amazon, suggested that most of the carbon taken up by Characiformes has its origin in the phytoplankton while the Siluriformes appear to acquire carbon from other plant sources.

In summary, most fishes of the Amazon appear to feed on a variety of plants and animals. In addition, the list of stomach contents for the main groups appears to be relatively uniform, an indication of the flexibility and the opportunistic character of the Amazonian fishes in their diets. We suppose that these two aspects have enabled many different species and a large number of individuals to share the same habitat in the Amazon.

3.2.2 When and How Much Is Eaten?

The availability of different food items in the Amazon basin is a dynamic process, it changes quality and quantity in response to the oscillations of the water level. During the seasons with high water levels, forest products (fruits, seeds, arthropods, flowers, etc.) are easily obtained. The great quantity of these products, relative to the amount of floating aquatic macrophytes, makes them preferential items for most fishes during this period. There are several examples illustrating this point. *Colossoma macropomum*, known as a fruit and seed eater, is one of them. Goulding, (1980) studying this species in the Madeira river, showed that during periods of high water levels, 94% of total volume of food ingested was fruits and seeds (from at least 13 different sources) and only 6% was of animal origin (fish and feces). During seasons when the water levels were low, he observed a reverse situation: 90% of the total volume of food ingested was of animal origin (fish, zooplankton, mayfly larvae, and cockroaches) while only 10% was fruits and seeds. *Piaractus brachypomum*, a close relative of *C. macropomum* was studied by the same author and showed a similar trend during seasons with high levels of water. However, during times when the water level was low, it fed mainly on leaves, about 81%, which represented only a small portion of the total volume (4%) ingested during the preceding season (see Fig. 3.5). Similar results were revealed by Santos (1991), while studying fish from Jamari, Machado, Guaporé, Mamoré, and Pacaás Novos rivers, in Rondônia, by Leite (1987), while studying fish from Uatumã river, and by Zuanon (1990), while studying *Lithodoras dorsalis* from Marchantaria island.

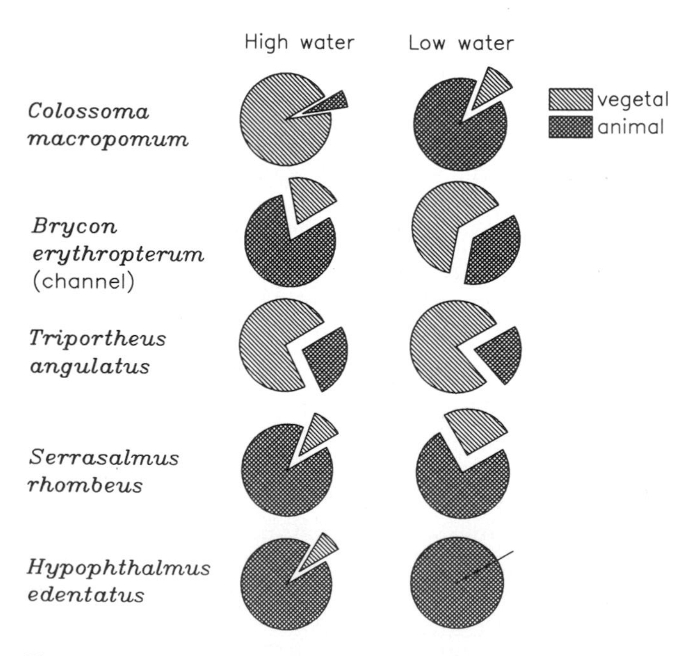

Fig. 3.5. Proportion of vegetal and animal material in the stomach of selected fish species of the Amazon during high and low water seasons. (Data compiled from Carvalho 1980; Goulding 1980; Santos 1981; Borges 1986; Leite 1987)

In addition, the mean bulk of stomach contents observed for almost all fish species, which are mostly or strictly herbivores, is always higher than 60% of the full capacity during high water and is about 5% or less during seasons when water levels are low. Thus, those fish species feeding on forest products eat as much as they can during seasons with high water levels in order to compensate for the off season when food is scarce.

Carnivores, on the other hand, present a reverse situation. After seasons of high water levels, the water retreats from the *várzea* and the lakes, causing a significant increase in the density of the ichthyofauna, and thus, food for the carnivores is plentiful and easy to procure. It is, therefore, reasonable that fish appear to acquire food mainly when the water levels are low. For example, Zuanon (1990), studying the catfishes from Marchantaria island, showed that specimens of *Brachyplatystoma flavicans*, a piscivorous fish, had virtually empty stomachs during the periods when the water levels were high, while during periods when the water levels were low, they fed on fish from different species. Similar results were shown by Leite (1987) who studied *Sorubim lima* from the Uatumã river. This species appears to feed mainly on fish when the water level is low while paelomid prawns and scales appear to be predominantly eaten during seasons when the water level is high.

Floating macrophytes, roots, tree branches, leaves, fruits, etc. are, as emphasized before, the main sources of detritus in the Amazon basin. As the water retreats, most of these materials are accumulated in the small lakes and close to the margins of the rivers providing food for detritivorous fish. Santos (1991), studying feeding behaviour of some species in the Jamari basin, observed a significant increase in the number of curimatids in the lake during low water season.

Almost all interactions among living beings and between them and the environment in the Amazon basin are shaped by water level oscillations. Since water level oscillations are not a synchronous event in the whole basin (see Chap. 2), these interactions are transient, i.e., they are different at different places at the same time. During unfavourable periods (low water season for herbivores or high water season for carnivores, for example), when food may become scarce, the energy needed for activities seem to rely on the flexibility of the animals in their diets and on the fat reserves built up in the muscle, liver, and abdominal cavity during the favourable periods. So, in summary, what, when, and how much is eaten by most fishes of the Amazon are mainly regulated by water level oscillations.

3.3 Migratory Cycles

Migration, as defined by Northcote (1984), is a periodic movement of a significant fraction of the population which results in an alternation of their habitats. In fact, this concept adds some restrictions to Baker's (1978) definition, which defined migration as "the act of moving from one spatial unit to another". In other words, fish migration, according Northcote's concept, does not include irregular (nonperiodic) movements or movements of small groups of fish, which actually appears to more appropriately describe the general movements of fishes of the Amazon. In addition, there are two different types of studies of fish migrations: those designed to describe the movements and those that attempt to explain them. Different methods have been used to study these aspects. For example, marking and tagging experiments, echo surveys, fishery statistics, and direct observation are suitable to describe fish migrations while behavioural studies in the laboratory are used to explain them (Jones 1968; Northcote 1984; Johnsen 1984; Smith 1985).

The study of fish migration in a complex environment such as the Amazon basin which includes thousands of different types of waterbodies, a significant and regular water level oscillation, and the appearance and disappearance of many aquatic habitats (see Chap. 2), is not an easy task. First, it should be emphasized that to this date nothing is known about the physiological control of fish migrations in the Amazon and only a few studies have tried to explain them. Most of what we know are descriptions of movements of a few species (most of which have commercial importance) based mainly on field observations and fisheries statistics. This is because marking and tagging experiments have failed in the Amazon (Godoy 1979; Goulding 1980; Worthmann 1982).

From these studies it was possible to identify at least three different types of potamodromous fish migrations in the Amazon basin: migrations due to changing water level, spawning migrations, and feeding migrations. No studies have been carried out on possible anadromous, catadromous, or amphidromous fishes in the estuary region of the Amazon, around the Marajó island, although they seem to occur.

Migrations due to changing water level are those relocations in a small space, particularly within the floodplain areas, when almost all individuals of some species move a few hundred metres or a few kilometres on a regular basis (Junk 1984b). As the water level rises, new feeding and hiding places become accessible. Although some environmental constraints (low O_2, for example) may occur in these areas, many fish species enter these places (floodplain lakes, flooded forest, etc.) at this time. When the water levels begin to fall again, many fish species migrate back to the main rivers, whereas others remain in the floodplain areas. The latter group includes many species resistant to hypoxic conditions (Junk et al. 1983; Val 1986; Almeida-Val et al. 1993; Soares 1993).

Spawning and feeding migrations described for fish of the Amazon basin include lateral and longitudinal movements. Lateral movements include those between floodplain lakes and rivers, while longitudinal movements include those directed upstream in the main rivers and those directed from headwaters of "black" or "clear" water rivers to "white" water rivers. While water level oscillations seem to be the main factor, lunar cycles, rain, and other environmental factors, appear to affect fish movements in the Amazon (Goulding 1980; Ribeiro 1983; Junk 1984b; Santos et al. 1991).

Movements of fish to breed and to feed between floodplain lakes and rivers have been observed for many fish species. Cox-Fernandes (1989) reported that some curimatide (*Potamorhina latior*, *Psectrogaster amazonica*, *Curimata knerii*), anostomids (*Leporinus trifasciatus*), and prochilodontids (*Prochilodus cf nigricans*), among others, migrate from Lago do Rei to the Amazon river as the water level begins to fall. In addition, she observed that after the lateral movements some schools migrate longitudinally in the main river. These movements are not motivated for spawning, since the gonad maturation process is not complete. As the water begins to rise, another movement takes place from the lake to the river, this time clearly for spawning purposes. Although some species spawn in the connecting canal (Petry 1989) most of them migrate upstream in the main river to spawn in other areas.

While migration between floodplain lakes and rivers covers a few hundred metres or a few kilometres, those directed from the headwaters of "black" and "clear" water rivers to "white" water rivers cover distances ranging from several hundred kilometers up to a few thousands kilometres. These types of migrations have been described for many characiforms and siluriforms based on field observations. Even though fish schools migrating near the water surface are easily observed, field observations require significant obstinacy, perseverance, and patience.

Many large characins (species of *Colossoma*, *Leporinus*, *Mylossoma*, *Prochilodus*, *Hemiodus*, *Anodus*, and *Semaprochilodus*, among others) from the Madeira/Machado rivers form schools and migrate (Goulding 1980). According to this author, from the beginning to the middle of the flood season, exact timing depending on the species, several schools of these species migrate downstream from the nutrient-poor tributaries to the main "white" water river where they spawn. After spawning they migrate back up the side tributaries and disperse across the flooded forest while the fertilized eggs remain in the nutrient-rich "white" water. The flooded forest, as discussed earlier, provides a temporary feeding area for the spent animals. When the water level begins to fall, the animals form schools again, migrate back to the main river and then upstream. The main tributary seems to provide refuge during seasons when the water levels are low.

Ribeiro (1983), studying the migrations of *Semaprochilodus* (jaraquis) through field observations, proposed an interesting model for migrations between "black" and "white" water rivers (Fig. 3.6). His model is similar to the general one proposed by Goulding (1980) for the animals from the Madeira/Machado system. Ribeiro thought *Semaprochilodus* species spawn in the confluence of "black" and "white" rivers, and then spent animals migrate back to black rivers, dispersing across the flooded forest. The genus *Semaprochilodus* includes two species, *S. insignis* and *S. taeniurus*, and a possible hybrid between the two (Vazzoler et al. 1983; Ribeiro 1985;

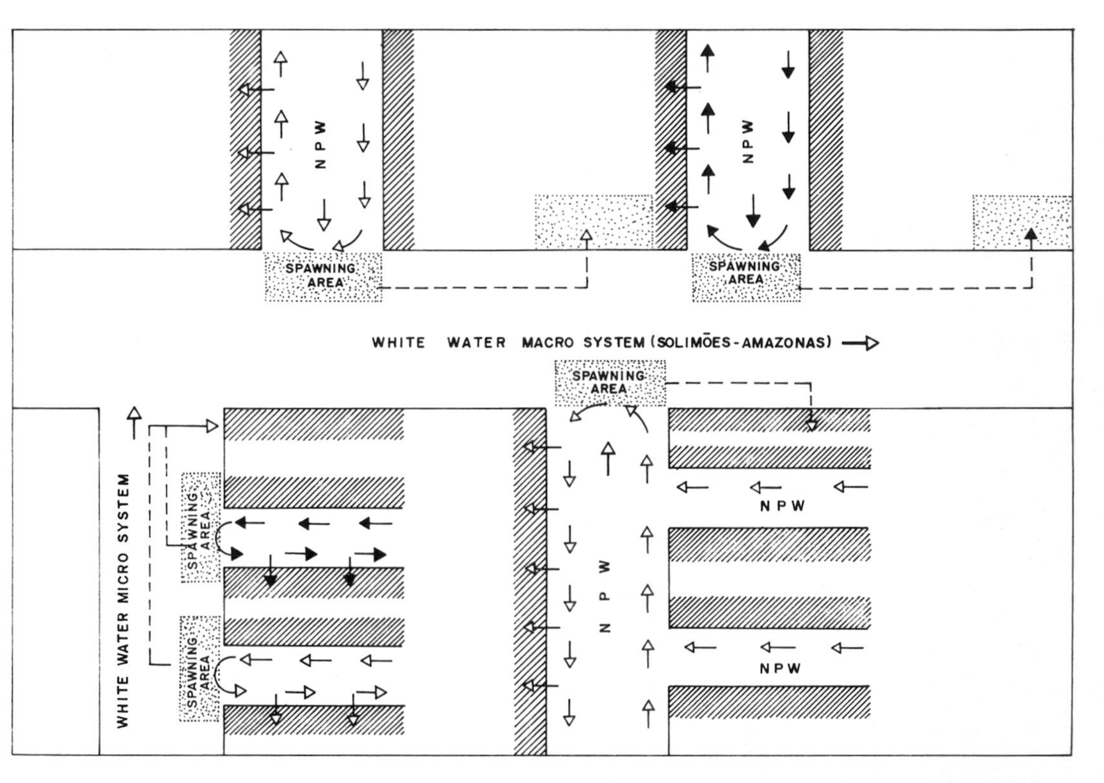

Fig. 3.6. Diagramatic representation of trophic and reproductive migrations of *Semaprochilodus* at the beginning of the flooding season. Note that the spawning areas are always located in the confluence of a "white" water system with nutrient-poor water (NPW) subsystems. (After Ribeiro 1983)

Feldberg et al. 1987). Both species have the same migratory behaviour, spawning at the same place. Although the spawning peaks are separated by an interval of a few weeks, some animals of both species spawn at the same time (Chaves and Vazzoler 1984; Vazzoler et al. 1989a,b).

The fertilized eggs are passively carried downstream in the "white" water river. Larval development takes place in the floodplain areas of this nutrient-rich system. As the water level begins to fall, these preadult animals, who have spent the past several months feeding, congregate into schools and migrate upstream through the "white" water river and then through the first "black" water tributary that they find. After a long migration, these animals disperse across the flooded forest, previously occupied by the adults, where they have plenty of food. Thus, there is a "compensation" for the displacement of young stages.

The movements of catfish of the Amazon are more difficult to detect than those of the characins because these animals swim near the bottom (Goulding 1980). However, Goulding (1979), who studied these animals as they passed through the rapids of the Madeira river, near Porto Velho city, reported that schools of 12 different catfish species were observed moving upstream through Teotonio's fall. Loubens and Aquim (1986) reported similar patterns for *Pseudoplatystoma tigrinum* (caparari) and *P. fasciatum* (sorubim) in the Mamoré basin, where these species migrate upstream towards the Andes to breed. In addition, Goulding (1989) reported that the preadult forms of *Brachyplatystoma flavicans* (dourada), recruited in the estuary region of the Amazon, migrate about 4000 km upstream through the Amazon river up to the Madeira basin. Except for these studies, there is virtually no information available describing or explaining catfish migrations in the Amazon.

3.4 Breeding Patterns

Fishes of the Amazon exhibit a profusion of breeding patterns including different sites, ages of first maturity, egg sizes, and care of eggs and fry. The studies of the reproductive characteristics of tropical fishes, in particular those focusing on Amazonian species, are heavily biased towards large species, despite the small adult size of the majority of tropical freshwater teleosts (Munro 1990). As previously described, many fish species migrate to reach an appropriate place to reproduce while others do not migrate. Among the migratory ones, two patterns are observed: almost all species spawn at the confluence of "black" and "clear" water rivers with "white" water rivers (Goulding 1980; Santos 1982; Ribeiro 1983; Araújo-Lima 1984; Zaniboni 1985; Correa 1987) while a few of them may spawn in the headwaters, like *Hypophthalmus* spp. in the Tocantins river (Carvalho and Merona 1986).

Most of the migratory fishes of the Amazon spawn once a year, normally just prior to or during the period of flooding. The spawning females belonging to this group are often highly fecund, shedding all of their eggs in a short period of time. In general, the migratory characins in the Amazon produce a large number of eggs and do not show any postfertilization care of the eggs (Lowe-McConnell 1987; Menezes and Vazzoler 1992; Vazzoler 1992). *Prochilodus*, whose ovaries account for up to a quarter of the total weight of the fish, shed about 300000 eggs (Schwassmann 1978) while *Colossoma macropomum* may shed as many as 500000 eggs (Machado-Allison 1987). The number of eggs shed each spawning season may vary from 36000 to 310000 in the *Brycon cephalus* (Zaniboni 1985). Each species, each population of fish has its own complex of adaptations that allow it to achieve reproductive success under specific environmental and physiological conditions. As mentioned above, the fertilized eggs are passively carried through the "white" water system reaching the floodplain areas where an abundance of food and refuges are available for the newly hatched fish. Included in this group are some species of Anostomidae, Serrasalmidae, Prochilodontidae, Bryconinae, and Triportheinae (Paixão 1980; Santos 1982; Zaniboni 1985; Petry 1989).

The nonmigratory fishes appear to spawn during all seasons indicating that water level changes have little effect on the reproductive process of these fish. Araújo-Lima (1984) reported that eggs and larvae of different species of characins are continuously observed over 10 months (from August to May) in the Solimões river near Manaus city. Also, in many places animals at different gonadal stages, including spent females, have been observed occurring simultaneously (Soares 1979). Other species clearly spawn during low water season such as *Osteoglossum bicirrhosum* (aruanã) and many loricariids (Aragão 1981; Brito 1981; Lowe-McConnell 1987; Machado-Allison 1987). Fishes inhabiting the small forest streams, where there are almost no appreciable seasonal changes in water level, exhibit partial spawning, like *Gymnorhamphichthys hypostomus*, whose reproductive activity consists of four successive spawning bouts (Schwassmann 1978). In these cases, environmental factors other than water level pulses and/or even behavioural characteristics appear to have a major role. These include lunar cycles, partial spawning, and parental care.

Many species of nonmigratory fishes of the Amazon spawn in lakes where environmental constraints are frequently present. To cope with these constraints some adaptive strategies have been used. So, some species like *Hoplosternum littorale*, *H. thoracatum*, and *Callichthys callichthys* form a group of floating air bubbles around their eggs in order to protect them from low O_2 levels (Lowe-McConnell 1987; Machado-Allison 1987). It is possible that the bubbles, by mimicking those produced by other animals (toads, for example), help in avoiding predation. Others, like *Serrasalmus* spp., *Pyrrhulina vittata*, *Hoplias malabaricus*, *Nannostomus* spp., *Cichla* spp., and

Geophagus spp., have adherent eggs which are placed on leaves and roots (reviewed by Santos 1991).

Parental care is exhibited by many species in different fashions. For example, *Loricaria* spp. and *Aspredo* spp. carry groups of 100 or 200 eggs glued together and attached to their bodies (Lowe-McConnell 1987). *Arapaima gigas* spawn between December and March and the male carries the fertilized eggs and the young offspring in the mouth (Menezes 1951; Salvo-Souza 1990). Nests containing fertilized eggs and young are guarded by many cichlids, for example *Cichla ocellaris*, *Cichlasoma festivum*, and *Astronotus ocellatus*. This behaviour is also exhibited by some serrasalmids, and by some catfish (Lowe-McConnell 1987; pers. observ.). *Astronotus ocellatus*, in addition, continuously mix the water close to the nest, apparently to maintain high levels of dissolved O_2 since the nests are often built in the floodplain areas during the low water level period (pers. observ.). How the young of these species are kept together is unknown; almost certainly pheromones are involved in this process, similar to that seen in *Arapaima gigas* (Lüling 1964).

Partial spawning behaviour has been described for some cichlids, *Hoplias malabaricus* and *Hoplerythrinus uniataeniatus*, and some gymnotiforms among others (Schwassmann 1978; Lowe-McConnell 1987; Santos 1991). Batches of eggs, ranging in quantity from a few hundred to several thousand, are shed each time. *Cichla ocellaris* has a minimum of 22 days between spawnings (Lowe-McConnell 1987) while other species spawn only twice a year. The environmental factor which drives the timing of the spawnings in these animals is unknown. Many native fishermen, however, believe that lunar cycles and rain might have some influence. In addition, it is possible that endogenous factors are involved, as seen in other fish groups (Stacey 1984).

The great majority of Amazonian fishes are oviparous with external fertilization. Some groups, however, exhibit different behaviours. For example, in Glandulocaudinae, who live mostly at the periphery of the Amazon basin, the fertilization occurs by means of a spermatophore because both sexes do not mature at the same time. To inject its spermatophore, males of this group appear to attract the female by a substance secreted by the caudal gland (Géry 1984; reviewed by Menezes and Vazzoler 1992). Internal fertilization has also been described for stingrays, guppies, and some catfish (Britsky 1972; Wourms 1981; Machodo-Allison 1987).

Fecundity, defined as the number of ripening eggs in the female prior to the next spawning period, varies significantly among the fishes of the Amazon. As a general rule, those species exhibiting parental care have a low fecundity. There is considerable variation, however, among each group. For example, among the guarders (fish guarding their eggs and young), the number of ripe ova in the ovaries is 180 for *Osteoglossum bicirrhosum* (Lowe-McConnell 1964; Aragão 1981), between 1000 and 3500 for

Astronotus ocellatus (Fontenele 1953), and about 47 000 for *Arapaima gigas* (Lowe-McConnell 1987). Among the fish that do not guard their eggs and young, the range is even greater, varying from less than 5000 up to 500 000 (Machado-Allison 1987).

According to Wootton (1990), the tendency in fish is to minimize egg size, thus maximizing the fecundity. However, as pointed out by the same author, the optimal egg size is that which maximizes the offspring survival rate. This seems to be the general rule among the fish of the Amazon since there is no clear relationship between phylogenetic position of the animal and the egg size for those species sharing the same environment. This analysis is possible in the Amazon because the evolution of the basin has resulted in a profusion of slightly different habitats which has allowed many fish species to become widely disseminated.

3.4.1 Larval Development

The survival rate of fertilized eggs and fry often depends on parental behaviour. In general, the higher the parental care the lower the mortality in early stages. Although parental care is exhibited by many fishes of the Amazon, the great majority of them lay a large quantity of eggs whose survival rate will depend almost entirely on environmental characteristics. Indeed, the synchronization between spawning time and some environmental characteristics plays an important role on the survival rates of eggs and fry. In addition, Araújo-Lima (1991), studying the larval development of 14 fish species in the Amazon, showed that the development of some traits (circulatory system and age at starvation, for example) is strongly related to the phylogenetic position of the species and their spawning sites. He also showed that cichlids are more resistant to starvation ($p = 0.008$) than are characids and catfish. Cox-Fernandes and Petry (1991) emphasized the importance of the floodplain areas in acting as a nursery for many non-guarding fish species in the Amazon. Because many fish species migrate as the water level begins to fall, the survival of young is vital in recolonizing the system. Otherwise, if for some reason a significant environmental disturbance occurs, an age class may be extremely reduced or even absent, as observed for *Plagioscion squamosissimus* in Janauacá lake (Worthmann 1982).

3.5 Communication

Communication seems to have an important adaptive value for fishes living in a complex system such as the Amazon. Except for the electrocommunication of some gymnotiforms and catfish, modes of communication have been poorly documented for fishes of the Amazon. These modes include changes in body colours and markings, sound, and production of very specific chemical substances (see Lowe-McConnell 1987 for examples of tropical fish). According to Kramer (1990) electroreception in teleosts has evolved at least twice: in some African Osteoglossomorpha and in the gymnotiforms and siluriforms groups. The electroreceptive notopterids and mormyrids do not occur in the Amazon, while siluriforms are well represented and the gymnotiforms are almost entirely restricted to the Amazon. For a review on electrocommunication in teleost fish, see Kramer (1990).

3.6 Evolutionary and Genetic Features

From an evolutionary point of view, fishes have been studied as a source of genome plasticity and experimentation, because they represent a large group of animals that can cope with hybridization more easily than other vertebrates (Whitt 1987). Furthermore, the diversity of their biological characteristics, as already stated for Amazonian fishes, makes this group fascinating for evolutionary studies.

The Amazon ichthyofauna not only provides a good example of the great diversification of fish, but also provides a good picture of the evolutionary process (see Table 3.1). The occurrence of extant species such as *Lepidosiren paradoxa* (lungfish), along with representatives of almost all existing orders of freshwater fish (including the most advanced teleosts), combine with the presence of many species of marine origin to provide biologists with an unexplored biological "gold mine". Furthermore, it has been reported that certain fishes in the Amazon can survive severe conditions such as lack of O_2, low pH, high sulphidric gas, and even large temperature drops. In addition, fish inhabiting both várzea and igapós areas must cope with several hours of anoxia in a single day and/or nearly the whole year with severe hypoxia (Junk et al. 1983; Almeida-Val 1986; Val 1986; Almeida-Val and Val 1990; Val et al. 1990; Almeida-Val et al. 1990, 1993). Increased speciation rates observed in fishes of the Amazon have been related to the high environmental heterogeneity. In addition, the uplift of the Andes further influenced these rates (see Baker 1970; Weitzman and Weitzman 1982 for details in this subject). Thus, the contemporary genome of the Amazon ichthyofauna may help in depicting the main geological, physiogeographical, and ecological events which occurred in the region.

3.6.1 Karyotypes of Fishes of the Amazon

The gene pool of a species or population includes all of the genes present in that species or population (Mayr 1977). The time factor is important because there have been many descriptions of genome changes and/or gene mutation events. These events per se represent the evolutionary history of species and populations and should never be ignored.

The vertebrate karyotypic evolution occurred through a series of genome duplication which originated from a primitive vertebrate genome. Therefore, a series of polyploidization events have occurred allowing for the extensive vertebrate radiation present today (Ohno 1970). According to Sola et al. (1981), the common karyotype of fish consists of 48 acrocentric (V-shaped) chromosomes. Many authors have suggested that there is a karyotypic conservation among both phylogenetically related and unrelated groups of fishes (reviewed by Sola et al. 1981). However, Robertsonian rearrangements (chromosome fusions and fissions) have occurred extensively in fishes leading to large variations in chromosome number while the number of chromosome arms is kept quite constant (Thorgaard 1983).

Rearrangements other than Robertsonian ones (inversions, duplications, deletions, and translocations) do not modify chromosome numbers, thus these kinds of changes can only be detected when more accurate cytogenetic techniques are employed. The study of chromosome number and shape can therefore help in understanding evolutionary phylogenetic relationships among species, but their evolutionary features can be better investigated when chromosome banding is performed. For the past 10 years, the Amazon fish species have been studied cytogenetically and many of their karyotypic characteristics (number, shape, banding) have been described for several groups (reviewed by Almeida-Val et al. 1991b; Porto 1992).

Reviewing karyotypic patterns of 211 fish species from the Amazon, Porto et al. (1992) showed that the chromosome numbers of Amazonian fishes vary from 2n = 22, as seen in *Nannostomus unifasciatus*, to up to 2n = 134, as seen in *Corydoras aeneus*. However, at the family level this variation is significantly reduced. Most species in Anostomidae (Bertollo et al. 1980; Galetti et al. 1991), Curimatidae (Venere and Galetti 1989; Feldberg 1990; Feldberg et al. 1992), Prochilodontidae (Feldberg et al. 1987; Pauls and Bertollo 1990), Hemiodidae (Porto 1992), and Chilodidae (Cestari et al. 1990) have similar karyotype morphology (2n = 54 distributed in meta- and submetacentric chromosomes). Such homogeneity, however, is not the rule among the characiforms; Erythrinidae (Bertollo et al. 1980, 1983; Giuliano-Caetano and Bertollo 1988), Lebiasinidae (Scheel 1973), Characidae (Scheel 1973), and Serrasalmidae (Porto et al. 1991) showed different karyotypic morphologies, presenting different chromosome numbers and multiple nucleolar organizing regions (NORs).

The available data on Siluriformes are not very extensive but nevertheless suggest that there is a high chromosome diversity in this group (Hinegardner and Rosen 1972; Scheel 1973; Fennocchio and Bertollo 1987; Oliveira et al. 1990; Porto and Feldberg 1992; Turner et al. 1992). Most authors suggest that this enormous diversity is due to gene duplication and/or polyploidization events (Fig. 3.7, reviewed by Whitt 1981). In the Amazon, natural triploidy was also reported in two Amazon species: *Eigenmannia* sp. and

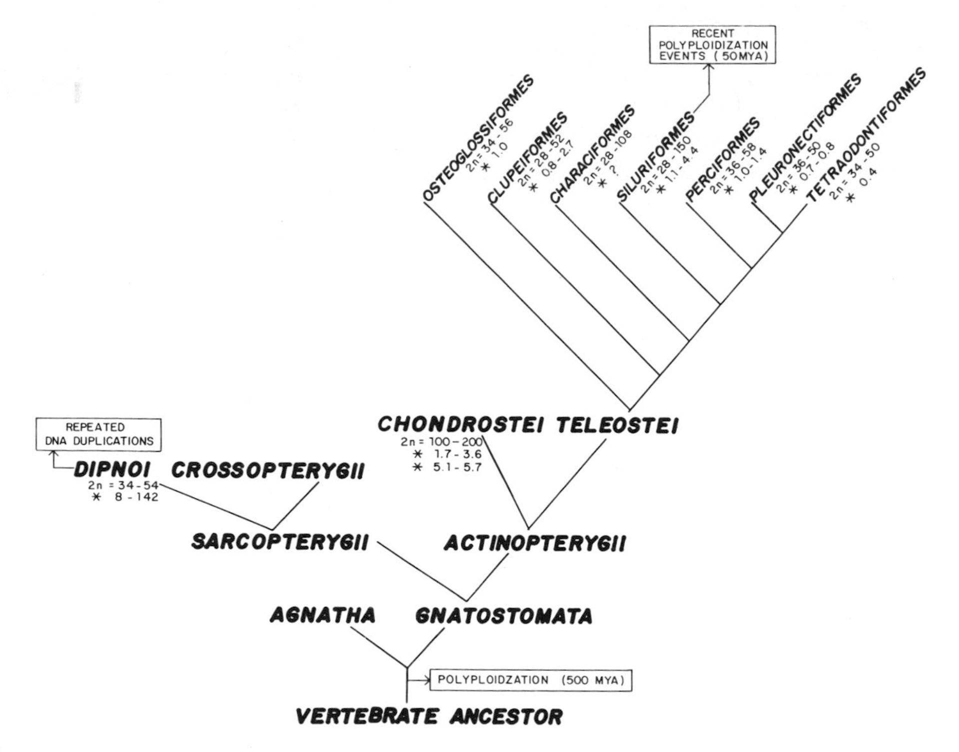

Fig. 3.7. Evolutionary relationships among main fish groups (according to Nelson 1984), showing the range of diploid numbers (zn) and the DNA contents (*, pg) for each group. Note the recent polyploidization events (50 Ma) in Siluriformes and the high level of DNA content in lungfish due to repeated regional duplications. (Data compiled from Ohno 1970; Hinegardner 1976; Kirpichnikov 1981; Whitt 1981; Almeida-Toledo et al. 1985; Giuliano-Caetano and Bertollo 1990)

Hoplerythrinus unitaeniatus (Almeida-Toledo et al. 1985; Giuliano-Caetano and Bertollo 1990).

The Perciformes is considered an advanced teleost group. Among them, the most studied group is Cichlidae. This family is characterized by a basic 2n = 48 chromosome number, most exist as subtelo-acrocentric units (Thompson 1979; Feldberg and Bertollo 1985). It is considered a very stable group although it does exhibit some variability such as 2n = 23 which is encountered in *Dicrossus filamentosus* (Thompson 1979) and 2n = 60 in *Symphysodon aequifasciatus* (Ohno and Atkin 1966; Thompson 1979).

In spite of this apparent stability, a more detailed examination of characiforms reveals the presence of sex chromosome heteromorphism, NOR transposition, and C-band (heterochromatin patterns) specific for each species. For example, sex chromosome differentiation was described for six Amazon species (2.8% of the total analyzed; reviewed by Porto et al. 1992).

At least three genera (*Triportheus*, *Semaprochilodus*, and *Eigenmannia*) include species in which the mechanism ZZ/ZW is observed (see Fig. 3.8). A more complex system was described by Bertollo et al. (1983) in the species *Hoplias malabaricus*. This species is polymorphic exhibiting different diploid numbers, different fundamental numbers, and XX/XY_1Y_2 formulae for its sex chromosomes. The presence of sexually differentiated chromosomes indicates an advanced stage in the genome evolution of the species.

Recently, Porto and Feldberg (1992) described the presence of quite different karyotype macrostructures and different NOR positions in two species of the same genus (*Hoplosternum*). These authors have suggested that these differences are related to the different geographic distributions and different population sizes in this group. White (1968, 1978) suggested that the presence of different karyotypes among closely related species can play an important role in evolution and speciation.

An interesting and controversial case of karyotypic evolution has been reported for species of Curimatidae. A low evolutionary rate in this family was proposed by Venere and Galetti (1989) based on similar chromosome numbers (2n = 54) for almost all analyzed species. However, the presence of different chromosome numbers (Scheel 1973; Feldberg 1990; Rocha and Giuliano-Caetano 1990) and different NOR distribution (Feldberg et al. 1992) in some genera of this family led Feldberg et al. (1992) to propose Robertsonian and non-Robertsonian rearrangements in the species of this

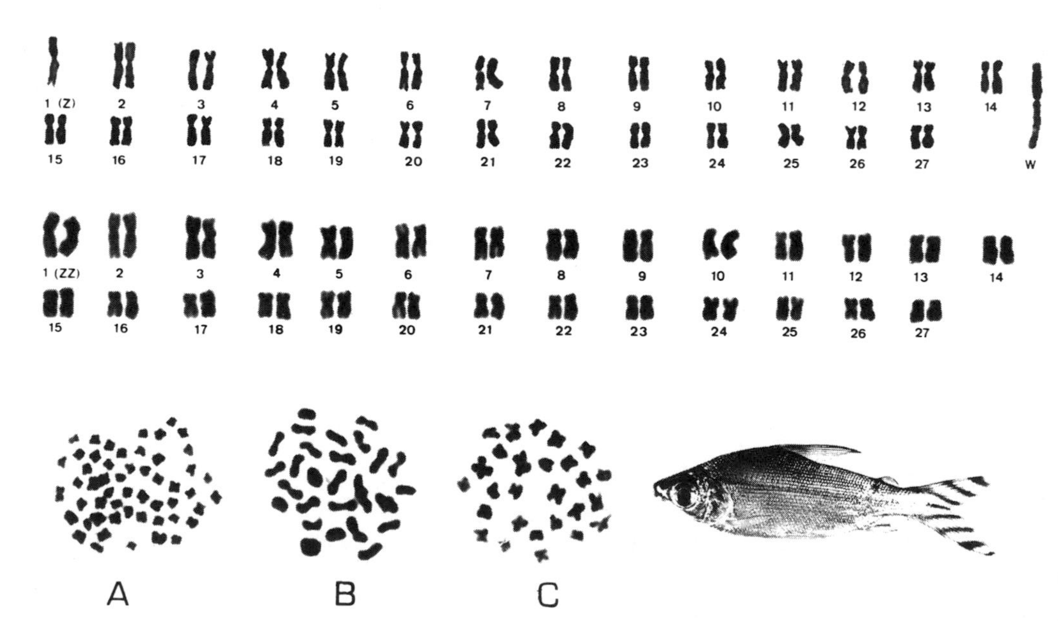

Fig. 3.8. Karyotype of *Semaprochilodus taeniurus* showing sex chromosome differentiation. Note the length of the W chromosome in the female. (Feldberg et al. 1987)

family. Feldberg et al. (1992) concluded that the evolutionary rates may not be as slow in this group as formerly proposed.

These characteristics confirm the evolutionary diversity among fishes of the Amazon who were formerly thought to be karyotypically stable (reviewed by Feldberg et al. 1992). Compared with fishes from other climatic regions (Ferguson and Allendorf 1991), this karyotypic heterogeneity, observed in such a small number of species from the Amazon, suggests that evolutionary experimentation is still occurring at relatively high rates.

3.6.2 Isozymes

Successive genome duplications during the first evolutionary radiation of vertebrates gave rise to many and multiple new types of proteins and therefore, new metabolic and adaptive opportunities (Ohno 1970; Whitt 1981). Gene duplication may also occur by regional duplication of specific DNA regions also resulting in the formation of different proteins. The duplicated gene may face different fates: it may remain similar to the original gene without any specialization, it may differentiate from the original gene, producing proteins for a specialized metabolic function (isozymes), or it may become silenced. The occurrence of true novelties, i.e., the creation of completely new genes, is an exceptional event since there is a strong tendency for the preservation of critical sequences of both structural and regulatory genes (Hochachka 1988a; Ferguson and Allendorf 1991). Thus, structural and functional properties of single enzymes may be helpful in evaluating evolutionary aspects of living beings. Isozymes are different forms of one enzyme which present the same specificity catalyzing the same reaction (Markert 1984). Only a few isozyme systems have been studied in Amazon fishes. The best analyzed example is lactate dehydrogenase (LDH), followed by malate dehydrogenase (MDH), and phosphoglucoisomerase (PGI) (reviewed by Almeida-Val et al. 1991b).

3.6.2.1 Lactate Dehydrogenase

The current distribution of lactate dehydrogenase LDH (L-lactate: NAD^+ oxireductase, E.C. 1.1.1.27) isozymes among teleost fishes is a product of the two forms of gene duplication mentioned above. LDH isozyme distribution in fishes of the Amazon follows the same basic pattern as that observed for teleosts in general: isozyme A_4 is expressed mainly in skeletal white muscle (primarily anaerobic tissue), but may be present in almost all tissues, even in highly aerobic ones; isozyme B_4 is expressed mainly in heart muscle (primarily aerobic tissue) and is present in several other tissues, sometimes predominating, sometimes not. This tissue distribution reflects functional characteristics of the isozymes that are well-established according to the preferential metabolism of each tissue. Thus, predominantly aerobic muscles contain a high proportion of B subunits (these subunits are effectively lactate oxidase) and typical anaerobic tissues may contain higher proportions of A subunits which are pyruvate reductase (Everse and Kaplan 1973; Hochachka 1980). Finally, isozyme C_4 Shows little tissue restriction in primitive teleosts and disappears in some intermediate groups. Its expression is drastically restricted in the advanced orders, occurring mainly as highly anodic isozymes in the eye and brain of Acanthopterygii fishes (reviewed by Almeida-Val and Val 1993) and as a cathodic isozyme in the liver of Gadiformes (this order of fish does not occur in the Amazon, Shaklee and Whitt 1981).

Some specialization of the isozyme B_4 has been detected in advanced teleosts and shows a highly restrictive tissue pattern. An example of this is the reduced expression of B_4 in heart tissue of flatfish (Markert and Holmes 1969) and stickleback (Rooney and Ferguson 1985). These fishes belong to the most specialized order of teleosts, the Pleuronectiformes. According to Kettler and Whitt (1986), the reduction in the expression of this isozyme in stickleback and flatfish tissues is related to the anaerobic conditions under which these fishes live. Our results from sampling LDH from Amazon cichlid fishes (Fig. 3.9) corroborate such tissue restrictive patterns. In addition, these fish species inhabit very hypoxic water and none of them have special structures for air-breathing or even for aquatic surface respiration.

According to Whitt (1983, 1987), the evolution of molecular and cell mechanisms induces differential expression of isozymes in tissues through the regulation of structural genes. Thus, nonspecialized enzymes would be expressed in a great number of tissues, and the most specialized ones would be characteristic of only a few tissues. A subsequent silencing process of a duplicated gene may be considered as one step in the evolution of species and/or genes (Ferris and Whitt 1979).

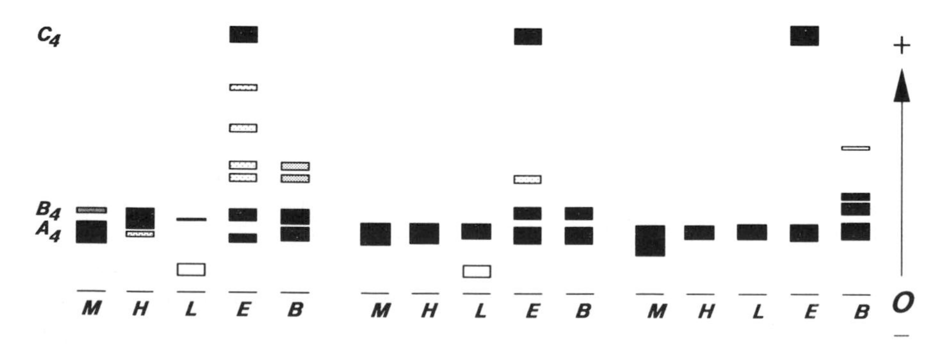

Fig. 3.9. Lactate dehydrogenase isozymes of representative species of cichlids of the Amazon. The expression of the homotetramer B_4 is restricted in almost all tissues, including heart. *M* Muscle; *H* heart; *L* liver; *E* eye; *B* brain

Distribution of LDH-C* product[1] throughout all tissues occurs in the species *Osteoglossum bicirrhosum* (D'Ávila-Limeira 1989) and *Arapaima gigas* (Almeida-Val, unpubl. data). On the other hand, this gene is not expressed in any of the tissues present in characins and catfish (D'Ávila-Limeira 1989; Almeida-Val et al. 1992), but is seen in some of the tissues of fishes belonging to the order Perciformes (superorder Acanthopterygii) as in the scianid *Plagioscion squamosissimus* (D'Ávila-Limeira 1989) and in several fish belonging to the group cichlids (Fig. 3.9).

The isozyme C_4 (LDH-C* gene) is not expressed in the liver or in neural tissues, such as retina and brain tissues, in Amazon characiforms and siluriforms (D'Avila-Limeira 1989; Almeida-Val et al. 1992). It has been suggested that C_4 isozymes, which are restricted to the eye in advanced teleosts, could be associated with the metabolism of photoreceptor cells in the retina of many teleosts, and play some role in the visual physiology of these fishes (Whitt et al. 1973; Whitt 1975; Coppes 1992). Panepucci et al. (1984), studying subtropical characins and silurids, described the presence of only two LDH loci (A and B) for all species. The Characiformes and Siluriformes belong to the superorder Ostariophysi and are the most numerous group of fishes in the Amazon (Table 3.1). As stated earlier, the Osteriophysi is a unique group of teleost fish in that they have a specialized auditory apparatus, the Weberian apparatus. If the LDH-C locus is involved with the visual system of teleosts, the successful appearance of another sensory organ, such as the Weberian apparatus, could have contributed to the diversification of the gene LDH-C* and further to the silencing of this specific gene in ostariophysans. A comparison of LDH characteristics of 52 fish species of the Amazon with fishes from other climatic regions is presented in Table 3.7.

Coexpressed eye- and liver-specific LDH isozymes have been detected in Amazon cichlids such as *Heros severum* and *Astronotus ocellatus*. This suggests that there is a fourth locus for LDH (Fig. 3.9). Leibel and Peairs (1990) described similar findings when studying the basketmouth cichlid *Acaronia nassa*. Further immunochemical analysis has confirmed a nonorthology of this liver-restricted LDH with LDH-C* loci detected in other species (Leibel et al. 1991).

In contrast to the five-banded pattern (Table 3.7) which is commonly seen in fish, three isozymes (A_4, A_2B_2, and B_4) are responsible for 30% of the LDH electrophoretic patterns observed for Amazon fishes. Two- or three-banded patterns are generally thought to be characteristic of phylogenetically advanced teleosts (reviewed by Almeida-Val and Val 1993). However, 12 species of Amazonian serrasalmids (piranhas and pacús) exhibit a high

[1] Nomenclature of enzyme subunits, structural gene or locus, and alleles is according to Shaklee et al. (1989).

Table 3.7. LDH loci, isozyme numbers (comprised of A and B subunits: IN) and C_4 isozyme tissue distribution (LDH-C_4/TD) in representative fish species of the Amazon

Order/species	LDH loci	IN	LDH-C_4/TD[a]	References[b]
Rajiformes				
Potamotrygon motoro	2	5	a	1
Osteoglossiformes				
Osteoglossum bicirrhosum	3	2	r/g	1/2
Arapaima gigas	3	2	g	3
Clupeiformes				
Pellona castelnaeana	3	2	g	1
Characiformes				
Brycon cf. *cephalus*	2	5	a	1
Brycon cf. *erythropterum*	2	5	a	1
Bryconops sp.	2	3	a	1
Raphiodon vulpinus	2	5	a	1
Roeboides myersi	2	5	a	1
Triportheus angulatus	2	4	a	1
Hoplerythrinus unitaeniatus	2	4	a	1
Hoplias malabaricus	2	3	a	1/4
Prochilodus nigricans	2	5	a	1
Hemiodus unimaculatus	2	5	a	1
Eigenmannina melanopogon	2	5	a	1
Potamorrhina latior	2	5	a	1
Potamorrhina altamazonica	2	5	a	1
Leporinus fasciatus	2	3	a	1
Leporinus agassizi	2	3	a	1
Mylossoma duriventris	2	3	a	5
Colossoma macropomum	2	3	a	5
Piaractus brachypomum	2	5	a	6
Myleus pacu	2	5	a	6
Myleus Prosumyleus spA.	2	5	a	6
Myleus rubripinis	2	5	a	6
Myleus schomburgkii	2	5	a	6
Mylesinus paraschomburgkii	2	5	a	1/6
Metynnis sp.	2	5	a	6
Catoprion mento	2	4	a	6
Serrasalmus Pristobrycon sp.	2	5	a	6
S. Pristobrycon striolatus	2	4/5	a	6
S. Pristobrycon eigenmanni	2	4	a	6
S. Pristobrycon hollandi	2	4	a	6
S. Serrasalmus sp.	2	4	a	6
S. Serrasalmus rhombeus	2	4	a	6
S. Pygocentrus nattereri	2	4/5	a	1/6
Utiarithchtys sp.	2	5	a	6
Siluriformes				
Brachyplatystoma filamentosum	2	5	a	1
Pseudoplatystoma tigrinum	2	5	a	1
Sorubim lima	2	3	a	1
Hoplosternum littorale	2	5	a	1/7

Table 3.7. *Continued*

Order/species	LDH loci	IN	LDH-C_4/TD[a]	References[b]
Perciformes				
Cichla ocellaris	3	2	hr	3/4
Heros severum	3	2	hr	3
Acarichthys heckelli	3	3	hr	3
Cichla monoculus	3	2	hr	3
Astronotus ocellatus	3	2	hr	3
Mesonauta insignis	3	2	hr	3
Plagioscium squamosissimus	3	3	hr	1
Pleuronectiformes				
Achirus achirus	3	2	hr	8

[a] g, general: isozyme C_4 appears in all or almost all tissues with similar intensities between both A_4 and B_4 paralogous isozymes; sr, short restricted: isozyme C_4 appears in several tissues but has some predominance in one or two tissues; r, restricted: isozyme C_4 appears in two or three tissues only and predominates in one of them; hr, highly restricted: isozyme C_4 appears only in the tissue in which it predominates; a, absent.
[b] (1) D'Avila-Limeira (1989); (2) Markert et al. (1975); (3) pers. observ; (4) Panepucci et al. (1984); (5) Almeida-Val et al. (1991a); (6) Almeida-Val et al. (1992); (7) Farias (1992); (8) Almeida-Val and Araújo (unpubl. data)

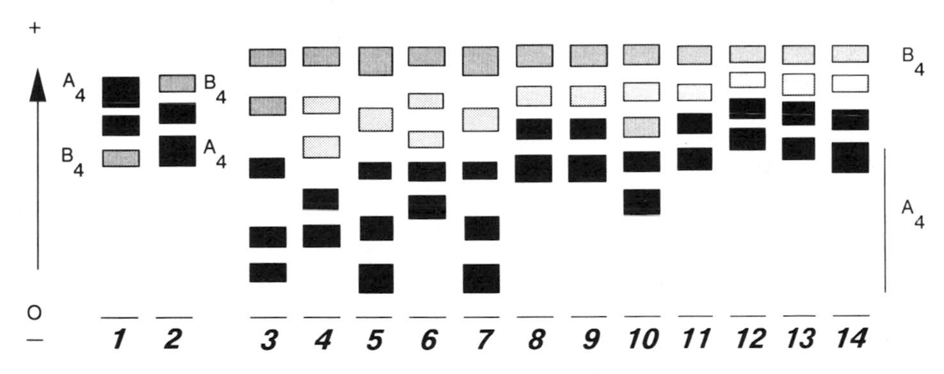

Fig. 3.10. Muscle lactate dehydrogenase isozymes of representative serrasalmids (piranhas and pacús) of the Amazon. An extensive heterogeneity of number of isozymes formed between A and B subunits is observed in this fish group. *1 Colossoma macropomum*; *2 Mylossoma duriventris*; *3 Myleus rubripinis*; *4 Metynnis* sp.; *5 Mylesinus paraschomburgkii*; *6 Myleus schomburgkii*; *7 Myleus* (*Prosumyleus*) sp.; *8 Serrasalmus* (*Pristobrycon*) *striolatus*; *9 S.* (*Pristobrycon*) *eigenmanni*; *10 S.* (*Pristobrycon*) sp.; *11 Serrasalmus* (*Serrasalmus*) *rhombeus*; *12 S.* (*Serrasalmus*) sp.; *13 Serrasalmus* (*Pygocentrus*) *nattereri*; and *14 Catoprion mento*

heterogeneity in the number of isozymes formed between A and B subunits (Fig. 3.10; Almeida-Val et al. 1992). Therefore, the generalization stated above, regarding isozyme number, should not be attributed to Amazonian fishes.

3.6.2.2 *Malate Dehydrogenase*

The enzyme malate dehydrogenase, MDH (L-malate: NAD^+ oxidoreductase, E.C. 1.1.1.37) is present in vertebrate cells in two different forms: mitochondrial (mMDH) and soluble (sMDH). The soluble form is composed of an isozyme system which is encoded at two loci (MDH-A* and MDH-B*). As a dimer, the combination of the two subunits results in a three-banded electrophoretic pattern (A_2, AB, and B_2). This pattern is present in almost all teleost groups (reviewed by Schwantes and Schwantes 1982) including most of the Amazonian fishes which have been analyzed (Almeida-Val et al. 1991b; Caraciolo et al., in press; pers. observ.).

Studies on Amazon cichlids have shown the occurrence of a six-banded rather than the regular three-banded electrophoretic pattern for sMDH in five different species: *Astronotus ocellatus*, *Cichla monoculus*, *Cichlassoma severum*, *Mesonauta insignis*, and *Geophagus* cf. *harreri* (Farias and Almeida-Val 1992). Those electrophoretic patterns can be explained as the result of the association between three subunits (A, B1, B2) which are encoded at the genes MDH-A*, MDH-B1*, and MDH-B2*, (Fig. 3.11). Six-banded patterns for MDH seem to be features restricted to the Amazon cichlids because cichlids from other climatic regions exhibit the common three-banded pattern (Schwantes and Schwantes 1982; Van der Bank et al. 1989).

These findings (i.e. the six-banded patterns) have been explained as the result of regional duplication ("in tandem") occurring at the MDH-B* locus

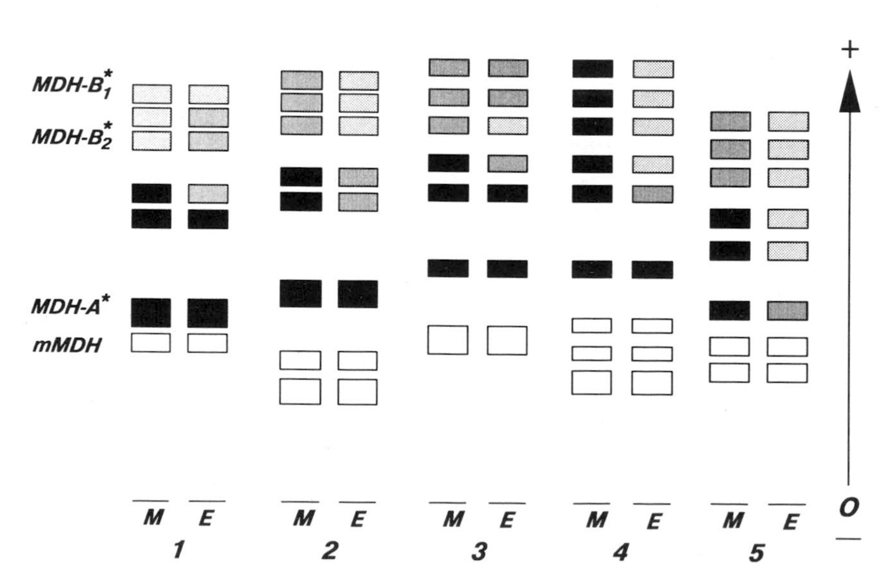

Fig. 3.11. Malate dehydrogenase isozymes of representative cichlids of the Amazon. *1 Astronotus ocellatus*; *2 Cichla monoculus*; *3 Mesonauta insignis*; *4 Cichlasoma severum*; *5 Geophagus harreri*. *M* Muscle; *E* eye. See Section 3.6.2.2 for details

in an ancestor of the Amazonian cichlids (Farias and Almeida-Val 1992). Preliminary experiments suggest that the duplicated loci have tissue distribution and restriction characteristics resembling those of the MDH-B* locus. In addition, thermostability and Klebe's tests have shown that the properties of the duplicated loci products are similar to those of the MDH-B* product. Thus, we suggest that this duplication occurred after differentiation between MDH-A* and MDH-B* loci and that the new duplicated genes undergo similar regulation (Farias and Almeida-Val 1992). Gene duplications are described by Ohno (1970) to be a result of polyploidy and/or regional duplication. Because there is no evidence that polyploidy occurred in Amazonian cichlids (Thompson 1979; Feldberg and Bertollo 1985), regional duplication appears to be the only plausible explanation. In addition, Whitt (1987) suggested that almost all isozymes in diploid fish arose from regional gene duplication.

Recently, Monteiro et al. (1991) reported a similar duplication of MDH-B* in the cichlid *Geophagus brasiliensis* found in the Paraná basin, southern Brasil. Their study showed that, out of 55 analyzed specimens, 48 displayed the same six-banded pattern that was described for Amazon cichlids. These authors also suggested that a duplication event at the MDH-B* locus could be responsible for their findings.

Based on fish species distribution, Géry (1984) suggested that the Amazon basin had recently been interconnected with neighbour basins and that the separation may still be incomplete (see Chap. 2). The ancestral species of the South American cichlids may have undergone several speciation events due to the different selective regimes which occured mainly after the uplift of the Andes. If there was a duplication event involving the MDH-B* locus in the ancestor of South America cichlids, the appearance of some three-banded specimens among the animals studied by Monteiro and coworkers may indicate the beginning of a silencing process in the Paraná basin cichlids. This could be attributed to environmental and biological differences faced by the animals living in subtropical (Paraná basin) and tropical regions (Amazon basin). In addition, the maintenance of the duplicated gene in the Amazon species could be explained as a heterosis phenomenon; the heterozygotes could have adequate selective power to spread out a recently introduced duplication (Farias and Almeida-Val 1992).

3.6.3 Allozymes

The origin of protein variability in vertebrates seems to be due mainly to successive polyploidization events in the ancestors of vertebrates (Ohno 1970) and, nowadays, to continuous genetic mutational events occurring in all living organisms (Nei 1987). As stated above, the isozyme systems most studied in Amazon fishes are LDH, MDH, and phosphoglucoisomerase, PGI (D-glucose-6-cetolisomerase, E.C. 5.3.1.9). Even though only a small number of species have been analyzed, allozymes appear to occur at different rates among these three systems (Table 3.8) with the occurrence of PGI alleles reaching values as high as 73% (Almeida-Val et al. 1991b). High allozymic levels of PGI are common in fish (Achaval 1986). However, MDH allozymes, occurring in 52% of the analyzed species, are an uncommon finding even though this enzyme system has been reported as very stable in most vertebrate groups, including fish (reviewed by Schwantes and Schwantes 1982).

Electrophoretic analysis of MDH in 12 species of Curimatidae showed that 6 out of 12 species presented similar alleles (B^*_{85}) for sMDH-B^* loci (Caraciolo 1989). Actually, similar electrophoretic mobility for both sMDH-A^* and B^* loci was observed for all species (Fig. 3.12) and so, the frequencies of B^*_{85} allele, which varied among the polymorphic species, appear to be the discriminating factor. The most parsimonious explanation is that a single mutational event occurred in the sMDH-B^* locus in the ancestor of these species (transpecific mutation) rather than being the result of many recent and independent mutational events. Since the frequencies for B^*_{85} allele ranged from 1.6 up to 43% among these species, it is possible that they have undergone different environmental or biological pressures after the mutation arose in the ancestor. Considering that the sMDH-B^*_{85} product is more thermostable than sMDH-B^*_{100} in five out of the six polymorphic species (Caraciolo 1989), this particular mutational event was probably advantageous during the evolutionary process of this family.

When compared with other species, the high frequencies of sMDH-B^*_{85} found in *Curimata ocellata* and *Curimata cyprinoides* (Table 3.9) suggest that different stages of allelic replacement are present. During this stage, the impact of drift is weak and the differences in the allele's adaptive values will be the major causes of their fixation or substitution (Kirpichnikov 1992). Systematic studies have shown that speciation events in this family precede the end of Andes uplift and, therefore, the Amazon basin formation (Vari 1989). The current status of sMDH in these species may be the result of modifications which occurred during the evolution of the basin per se.

Table 3.8. Occurrence of allozymes in fishes of the Amazon

Analyzed enzymes[a]	Number of species	Species with allozymes	Proportion of polymorphic species (%)
PGI	11	8	73
sMDH	29	15	52
LDH	41	7	18

[a] PGI, phosphoglucoisomerase; sMDH, soluble malate dehydrogenase; LDH, lactate dehydrogenase.

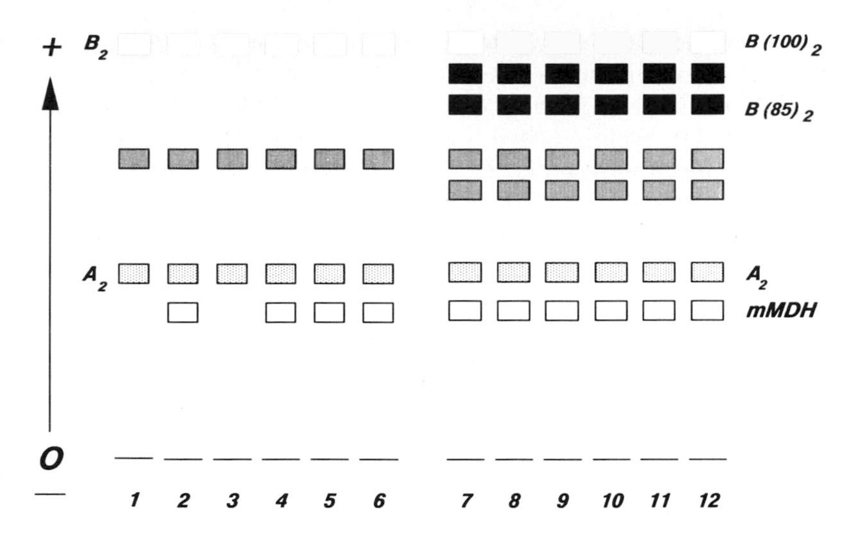

Fig. 3.12. Muscle malate dehydrogenase isozymes of representative curimatids of the Amazon. Note the conservative electrophoretic pattern for monomorphic (*1–6*) and for polymorphic (represented by heterozygote phenotypes) species (*7–12*). *1 Potamorhina latior*; *2 Curimata knerii*; *3 Curimata vittata*; *4 Psectrogaster amazonica*; *5 Curimatella alburna*; *6 Curimatella meyeri*; *7 Potamorhina pristigaster*; *8 Potamorhina altamazonica*; *9 Curimata ocellata*; *10 Curimata innornata*; *11 Curimata cyprinoides*; and *12 Psectrogaster rutiloides*

Table 3.9. Allelic frequencies of sMDH-B* (MDH-B_{100} and MDH-B_{85}) in 12 species of Amazonian curimatids

Species	p(B_{100})	q(B_{85})	n
Potamorhina pristigaster	0.7432	0.2568	37
Potamorhina altamazonica	0.9713	0.0287	87
Potamorhina latior	1.0000	0.0000	121
Curimata ocellata	0.6250	0.3750	8
Curimata vittata	1.0000	0.0000	7
Curimata knerii	1.0000	0.0000	7
Curimata cyprinoides	0.5714	0.4286	21
Curimata innornata	0.9348	0.0652	23
Psectrogaster amazonica	1.0000	0.0000	8
Psectrogaster rutiloides	0.9844	0.0156	96
Curimatella alburna	1.0000	0.0000	12
Curimatella meyeri	1.0000	0.0000	8

Nonrandom allelic distribution was described in another Amazon characin, *Leporinus friderici*, by Renno et al. (1990). This species is widely distributed throughout South America, occurring in rivers of the Guiana shield, in the Amazon basin, and in the Brasilian South basin of Paraná river (Garavello 1979). Renno et al. (1990) collected *Leporinus friderici* from several different locations; six different places in French Guiana and two in Brasil, including one in the Amazon basin. Using these samples, Renno and coworkers compared 21 loci from each fish; loci which produced a total of 53 alleles. Based on genetic markers (allelic distribution), two subpopulations were found in the Guiana rivers, one of them fairly closely related to the Brasilian populations. This distribution is coincident with the paleogeographic history of the populations and is explained based on the refuge theory, which supports the existence of microhabitats during the alternating glacial and interglacial ages. These refuges may have contributed greatly to the large organic diversity present in the Amazon today (Gentilli 1949; Vanzolini 1973; Ab'Saber 1977; Weitzman and Weitzman 1982). The results of Renno and coworkers show high values of population heterogeneity (H = 4–11%), typical of widespread fish species.

Leporinus agassizi and *Serrasalmus* (*Pristobrycon*) *striolatus* are also good examples of species having a high frequency of variation in the LDH-A* and LDH-B* loci (D'Ávila-Limeira 1989 and Almeida-Val et al. 1992b, respectively), thus exhibiting allelic fixation (Table 3.10). For *Metynnis* sp., the existence of two distinct patterns for sMDH-A* and PGI-B* suggests that this species is composed of two subpopulations (Fig. 3.13; Almeida-Val et al. 1991b). However, analyses of morphological and classic anatomical traits indicate that there is only one species (M. Jégu, pers. comm.). A similar situation has been observed in *Myleus rubripinis* (VMFA-V unpubl.

Table 3.10. Genotypic frequencies observed and expected for LDH loci of two species of Amazonian fishes

Leporinus agassizi				
LDH-A*	$A_{100/100}$	$A_{100/57}$	$A_{57/57}$	Total
Observed cases	9	8	6	23
Observed frequencies	0.39	0.35	0.26	1
Expected frequencies	0.32	0.49	0.19	1
		$pA_{100} = 0.565$		
		$qA_{57} = 0.435$		
LDH-B*	$B_{100/100}$	$B_{100/133}$	$B_{133/133}$	Total
Observed cases	7	10	6	23
Ovserved frequencies	0.30	0.44	0.26	1
Expected frequencies	0.27	0.50	0.23	1
		$pB_{100} = 0.520$		
		$qB_{133} = 0.480$		
Serrasalmus (*Pristobrycon*) *striolatus*				
LDH-A*	$A_{100/100}$	$A_{100/54}$	$A_{54/54}$	Total
Observed cases	9	8	2	19
Observed frequencies	0.47	0.42	0.11	1
Expected frequencies	0.46	0.44	0.10	1
		$pA_{100} = 0.680$		
		$qA_{54} = 0.320$		

data). Thus, it is possible that the diversity of Amazon fishes is greater than that estimated based solely on morphology and anatomy.

Morphological diversity and its evolution are caused mainly by requlatory gene changes (King and Wilson 1975; Wilson 1975). According to Nei (1987), regulatory genes are much more stable than structural genes. It is believed that the primary mechanism of biochemical adaptation and evolution is the selection pressure acting upon regulatory alleles (see Wilson 1985). Unfortunately most of the recent studies have dealt with biochemical diversity and variability of structural genes rather than with regulatory genes.

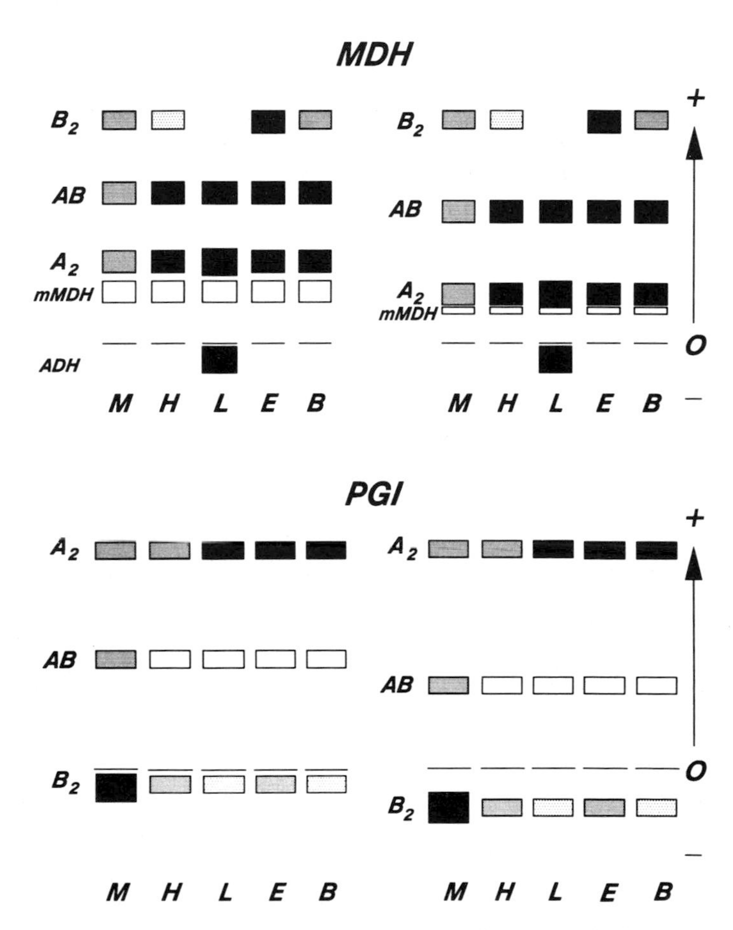

Fig. 3.13. Schematic representation of discontinuous phenotypes of malate dehydrogenase (*MDH*) and phosphoglucoisomerase (*PGI*) isozymes of *Metynnis* sp. *M* Muscle; *H* heart; *L* liver; *E* eye; *B* brain; *mMDH* mitochondrial MDH

3.6.4 Other Proteins

Many types of hemoglobin have been extensively studied in Amazonian fishes. Almost all species which have been analyzed present more than two electrophoretic hemoglobin fractions; only 5% of them have a single Hb component (Fyhn et al. 1979; Val et al. 1987). During the Research Vessel (R/V) Alpha Helix expedition to the Amazon, Fyhn et al. (1979), upon studying 96 species of fish through disc electrophoresis with polyacrylamide, reported no significant relationship between number of Hb fractions and phylogenetic status (Fig. 3.14). Some groups of fish, however, have a significantly different number of Hb fractions; for example, fishes from the superorder Ostariophysi presented a modal value of three bands, while those from the superorder Acanthopterygii exhibit a modal value of nine bands (Fig. 3.15; Fyhn et al. 1979). In addition, these authors observed a trend towards decreasing Hb multiplicity, especially for the major bands, with increasing specialization occurring in the characids and silurids (see Fig. 3.14).

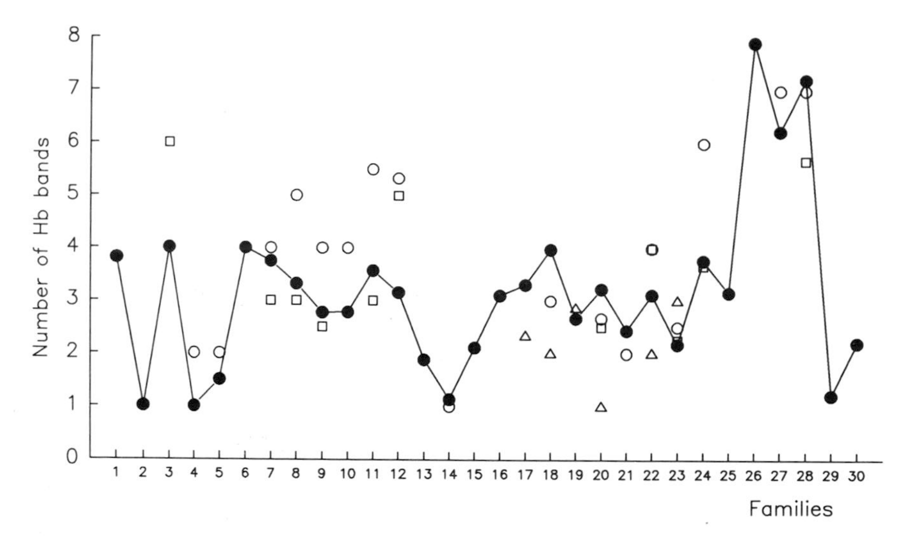

Fig. 3.14. Mean number of hemoglobin fractions of selected families of fishes of the Amazon basin. The families are organized along the abscissa from the generalized to the specialized according to Greenwood et al. (1966), Lauder and Liem (1983b), and Vari (1989). *1* Potamotrigonidae; *2* Lepidosirenidae; *3* Clupeidae; *4* Osteoglossidae; *5* Arapaimidae; *6* Characidae; *7* Serrasalmidae; *8* Erythrinidae; *9* Cynodontidae; *10* Prochilodontidae; *11* Curimatidae; *12* Anostomidae; *13* Hemiodontidae; *14* Gymnotidae; *15* Electrophoridae; *16* Apteronotidae; *17* Rhamphichthyidae; *18* Doradidae; *19* Auchenipteridae; *20* Pimelodidae; *21* Ageneiosidae; *22* Hypophthalmidae; *23* Callichthyidae; *24* Loridariidae; *25* Belonidae; *26* Synbranchidae; *27* Scianidae; *28* Cichlidae; *29* Soleidae; *30* Tetraodontidae. Data compiled from: *closed symbols* (Fyhn et al. 1979); *open triangles* (Galdames-Portus et al. 1982); *open squares*, (Perez and Rylander 1985), freshwater fishes collected in eastern Venezuela, using polyacrylamide gels; and *open circles* (Val et al. 1987), using agar-starch gels

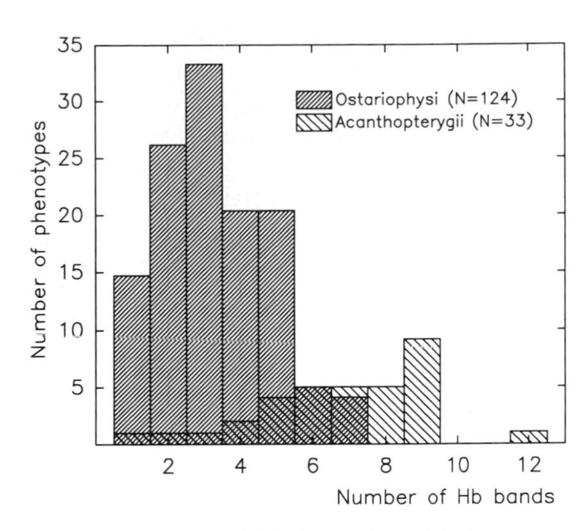

Fig. 3.15. Number of phenotypes per number of hemoglobin bands in ostariophysan and acanthopterygian fishes of the Amazon. (Redrawn from Fyhn et al. 1979)

Val et al. (1987), studying 22 species which were collected close to the Marchantaria island, reported similar trends for Hb distribution among the main groups. Mean Hb bands, however, should not be considered as a strong basis for the classification of groups because they vary depending on the species analyzed and on the electrophoretic support. Fyhn et al. (1979) and Galdames-Portus et al. (1982) reported a similar number of Hb bands (3.2 and 2.52, respectively) in Amazonian silurids in spite of the occurrence of highly banded patterns such as those found in *Pterygoplichthys multiradiatus* (Val et al. 1987, 1990). With the exception of this species, the average (2.5) reported for the group by Val et al. (1987) is similar to that observed by the previous authors.

More than one phenotype was described for several Amazonian species by Fyhn et al. (1979) during their survey. For example, the species *Mylossoma* sp. and *Prochilodus* sp. contain two and three plus one variant phenotypes, respectively. Further studies (Val et al. 1986, 1987) showed that the different phenotypes occurred due to the existence of more than one species of the genus. Discussing their results, Fyhn et al. (1979) pointed out that the difficult taxonomy associated with Amazonian fishes was a complicating factor when interpreting their results. The family Prochilodontidae includes different species belonging to different genera, i.e., *Prochilodus* cf. *nigricans*, *Semaprochilodus insignis*, and *S. taeniurus*. In addition, there is a possible hybrid between the two species of *Semaprochilodus* (Ribeiro 1983; Feldberg et al. 1987) that might have further complicated Fyhn and coworkers' analysis. Val (1983) and Val et al. (1986) reported three different hemoglobin phenotypes for the species of *Semaprochilodus*. These findings further support the presence of two species and a possible hybrid.

Because such Hb multiplicity has been shown, in many species, to have different functional characteristics, it can be assumed that this attribute in fish is an adaptive property (Garlick et al. 1979; Riggs 1979; Powers 1980). However, the biological significance of this multiplicity has not been ascertained yet because no clear relation exists between number of Hb bands and environmental stability. For example, Fyhn et al. (1979) pointed out that the Hb heterogeneity observed in Amazonian fishes is not related to different habitats or biological characteristics of the analyzed species. In addition, the surveys made by Galdames-Portus et al. (1982), Perez and Rylander (1985), and Val et al. (1987), when taken together, clearly show that the number of Hb bands is not related to environmental characteristics.

Nevertheless, the above-mentioned studies have not included any evaluation of the effects of seasonal environmental changes on Hb patterns. These changes, for at least a few species, seem to have a regulatory effect on the concentration of the bands rather than on their number (Val 1983, 1986; Val et al. 1986; Affonso 1990; Ramirez-Gil 1993). We conclude that evolutionary and environmental pressures have had only a minor effect on the number of Hb fractions present in fish. This, however, does not exclude the significant effect of evolutionary and environmental pressures in modelling the structure of the hemoglobins. Again, fishes of the Amazon provide an interesting biological example.

Reichlin and Davis (1979) analyzed the extensive cross-reactions exhibited by hemoglobins of 31 fish species of the Amazon to carp and trout Hb (I and IV) antisera and described some striking observations. First, no reactions were observed between the hemoglobins of *Lepidosiren paradoxa* and all three types of antisera, suggesting that the surface of the hemoglobins of *L. paradoxa* is very distantly related to trout I and IV and carp hemoglobins. In fact, as pointed out by Fink and Fink (1979), *Lepidosiren* "in terms of genealogical relationships is more closely related to *Homo sapiens* than to any other living fish with the exception of *Latimeria*, the coelacanth". Secondly, no fish hemoglobins precipitated with antiserum from trout HbIV (anodal component) while almost all of them precipitated with either antiserum to trout HbI (cathodal component) and to carp hemoglobin. This is a clear indication that the surface of trout HbIV and the analyzed Amazonian fish Hb are very distantly related.

Four patterns were observed for the hemoglobins that exhibited precipitation reactions to anti-trout HbI and carp Hb. Pattern A includes those reacting only with anti-carp Hb as seen in *Arapaima gigas*; pattern B is represented mainly by characids where the anodal components react more strongly with anti-carp Hb than with anti-trout HbI; pattern C includes those reacting equally to both antisera such as displayed by *Rhamphichthys* sp.; and finally pattern D includes those reacting almost exclusively with anti-trout HbI as seen in *Pterygoplichthys* sp. Their observations about the hemoglobins of catfish are very interesting. It is clearly shown that the hemoglobins of air-breathing species react strongly with anti-trout HbI serum, while those of water-breathing strongly reacted with anti-carp serum.

Similar general views corroborating the existence of an enormous inter- and intraspecific protein heterogeneity has also been drawn from studies dealing with other protein systems such as transferrins and crystalline proteins (Vazzoler 1983; Teixeira and Jamieson 1985). Thus, the general diversity of fish living in the Amazon, which has been translated into the number of fish species (estimated between 1300 and 3000 living species, considering new descriptions), does not include the heterogeneity of structural and regulatory genes. This heterogeneity, hidden in the nucleus of each cell of each living fish, seems to be particularly important since it provides the vital adaptive responses to the environmental and biological constraints faced by the organisms living in an environment such as the Amazon.

Chapter 4

Gas Exchange

Air and water are very different as a medium for gas exchange. Except for few extreme conditions, for example, animals living at high altitude and in secluded environments, the gas composition of ambient air does not fall below the acceptable levels for living beings. This is due to the enormous mass of air, subjected to powerful convective currents, which acts as a buffer for local variations. Even some environmental pressures generated by modern civilization (high production of CO_2, pollution, among others) have had, up to the present, no noticeable effect on gas exchange processes (Boutilier 1990; West 1991; Hochachka et al. 1991).

The composition of dissolved gas, on the other hand, is influenced by many different processes, including physical, chemical, and biological processes. Some of these influences on the Amazonian aquatic environments have been discussed earlier (in Chap. 2). Compared to temperate ecosystems, a peculiar characteristic of some water tracts in the Amazon is the intense diurnal fluctuations of O_2 and CO_2 in addition to the seasonal changes due to water level oscillation. Diurnal changes in the Amazon are significantly affected by respiration and photosynthesis. During daylight hours, photosynthesis produces high levels of O_2, surpassing the O_2 demand of the system. Deficiency of light, on the other hand, induces a decrease in O_2 production and the water becomes hypoxic or even anoxic. As the water level rises there is a dilution effect and the system may become very hypoxic (see Chap. 2). In addition, extreme hypoxia, as is sometimes observed in Amazon waters, seems rare in most temperate systems. For example, Davis (1975), basing his conclusions on a large number of experimental studies, suggested that 3.98 mg O_2/l of water would induce symptoms of O_2 distress in the average member of a temperate fish community. This level is higher when the salmonids are included. Oxygen levels are often much less than 3.98 mg O_2/l in the Amazon, in other words, hypoxia is widespread throughout the Amazon compared to temperate water systems. It would be erroneous to conclude that extremely low values of dissolved O_2 are representative of all Amazonian waters. Open areas, moving water, reduction of macrophyte cover, and wind, among other factors, may ameliorate the O_2 availability (see Chap. 2).

The explosive radiation of fish species facing similar environmental pressures, such as those observed in the Amazon, gave rise to an equally diverse range of adaptive solutions. These adaptive solutions occur at all biological levels, including behavioural, physiological, biochemical, and genetic levels. Slobodkin and Rapoport (1974) claimed that periodic environmental perturbations which are similar in nature can cause genetic modifications which result in a spectrum of adaptive solutions. In addition, multiple adaptive responses, i.e., involving more than one solution, are very often observed. For example, O_2 demand may be paralleled during hypoxia by improving O_2 transfer and/or by controlling energy expenditure. Similarly, O_2 transfer can be improved by many parallel solutions.

Some of these aspects regarding Amazonian habitats and their fishes were recognized by Hochachka and Randall (1978) when leading the Alpha-Helix expedition in the Amazon. They pointed out: "The Amazon was chosen as a study area because of an abundance of air-breathing fishes. However, early in the program our observations indicated that to breathe air is not the only solution to the problem of hypoxic Amazonian waters. Nor is hypoxia the only difficulty faced by these fishes. Large daily oscillations in temperature, O_2 and CO_2 levels, and pH, extremely low ionic content, and seasonal drought are equally formidable challenges . . . ." Many adaptive solutions adopted by these fascinating Amazonian animals have been described. They have greatly expanded our understanding of the evolutionary processes and adaptive mechanisms which have resulted due to the prevailing environmental conditions in the Amazon (Junk et al. 1983; Almeida-Val et al. 1993; Val 1993).

4.1 The Obligatory Air-Breathers

Obligatory air-breathers refers herein to fish species that rely primarily on the O_2 uptaken from the air. Since the Cambrian period, hypoxia and anoxia have been common events in the evolutionary history of waters (Randall et al. 1981). Even today many aquatic systems exhibit periodic or regular episodes of hypoxia or anoxia. These types of fluctuations are seen on the extensive floodplains in the Amazon. The development of air-breathing systems may have arisen accidentally in fishes that were skimming water surfaces during the periods of low O_2 (Gans 1970). The transition from water- to air-breathing was undoubtedly an important transformation during the evolutionary process of vertebrates, especially when considering the colonization of terrestrial environments by aquatic animals. However, unlike the extensive information on morphological aspects of extant air-breathers, there is but scant physiological and biochemical information on gas exchange in ancient fishes (Randall et al. 1981; Burggren et al. 1985; Almeida-Val and Hochachaka, in press).

The utilization of air as a source of O_2 has many advantages for fish, including independence from the fluctuations in dissolved O_2, reduced work in pumping an O_2-rich fluid, and decreased size of the pump needed (Gans

1970). Of course, there are also some disadvantages, including disturbance of the hydrostatic balance mechanisms in fish (Gans 1970), exposure to aerial predation, and exposure to frequent temperature changes. Many attempts were made during the evolutionary process by many different species of fish to solve the problem of tissue oxygenation during periods of hypoxia. Air-breathing is one solution and so, it was tried independently many times during the explosive radiation of fishes. As a consequence, a profusion of different structures associated with O_2 uptake from air have been described among living fish of different groups. In the Amazon, for example, at least nine families have air-breathing representatives (Table 4.1) and it is quite likely that they have independently evolved this capacity (Kramer et al. 1978).

Air-breathing fishes, however, represent only a small fraction of the living fishes, even in waters with low levels of O_2. Carter and Beadle (1931a), for example, reported that only 8 of 20 species that were analyzed during their classical expedition to the Paraguayan Chaco were capable of air-breathing.

Table 4.1. Structures associated with O_2 uptake from air in fishes of the Amazon. The familes are organized from the generalized to the specialized according to Nelson (1984)

Families and species	Structures			
	SL	S	SI	PBM
Lepidosirenidae				
Lepidosiren paradoxa	X			
Arapaimidae				
Arapaima gigas	X			
Erythrinidae				
Erythrinus erythrinus	X			
Hoplerythrinus unitaeniatus	X	X	X	
Doradidae				
Doras			X	
Callichthyidae				
Callichthys			X	
Hoplosternum			X	
Loricariidae				
Plecostomus			X	
Ancistrus			X	
Rhamphichthyidae				
Hypopomus				X
Electrophoridae				
Electrophorus				X
Synbranchidae				
Synbranchus marmoratus				X

[a] SL, Swim bladder and lung; SI, stomach and intestine; PBM, pharyngeal, branchial and mouth diverticula; and S, skin.

Similar observations were reported by Beebe (1945) for fishes of a muddy pond in Venezuela, by Kramer and Graham (1976) studying the fishes of a lowland stream in Panama, and by Junk et al. (1983) for fishes found in the Camaleão lake in the Amazon. Dissolved O_2 in all these locations is very low. In addition, the environmental conditions found in the Paraguayan Chaco, described by Carter and Beadle (1931b), are very similar to that of the Amazon floodplain areas. Interestingly, all air-breathing fishes reported by those authors (Carter and Beadle 1931a) are often found in the Amazon.

Although the number of air-breathing fishes is very small when compared with other fish species that share the same environmental conditions, it seems clear that waters which are naturally poor in O_2 are one of the key factors that induced the evolution of air-breathing in fishes (Packard 1974). In fact, the number of air-breathing fish species living outside the tropics, in normoxic waters, is proportionally reduced. The relationship between occurrence of air-breathing fishes and hypoxic waters has been reviewed by many authors, including Carter and Beadle (1931a) based on their own findings during the expedition to Brazil and Paraguay in 1926/1927; Kramer et al. (1978) who discussed the number of air-breathing fish found during the Research Vessel Alpha Helix expedition to the Amazon, Randall et al. (1981) and Burgreen et al. (1985) who discussed the evolution of respiratory aspects of air-breathing fishes.

It has been suggested that air-breathing vertebrates have evolved from Osteichthyes (see Randall et al. 1981). Chondrichthyes, both Holocephali and Elasmobranchii, do not have any representatives who possess morphological adaptations for breathing air. However, air sacs were present in the placoderm *Bothriolepis* (Denison 1941). Gans (1970) suggested that the primary function of these earliest air sacs was respiratory, but Randall et al. (1981) viewed this placoderm as a possible exception. As a consequence, two phylogenetic diagrams emerge: one that includes the Placoderms as having the primary "lungs" (Gans 1970) and the other that does not include this group of fish (Randall et al. 1981). Whether or not we include the Placoderm group, all living air-breathers seem to have evolved from the same group of bony fishes which were distributed throughout a similar range of environments and were exposed to similar environmental constraints.

According to Nelson (1984), the Osteichthyes includes four groups: Dipneusti, Crossopterygii, Brachiopterygii, and Actinopterygii. The Dipneusti, or true lungfish, includes the order Lepidosireniformes, which has one representative living in the Amazon, *Lepidosiren paradoxa*. The Crossopterygii is now extinct, except for the strictly aquatic coelacanth *Latimeria chalumnae*. The Brachiopterygii includes the family Polypteridae which has ten species which occur in Africa. The Actinoperygii includes Chondrostei and Neopterygii, both having representatives which are either strictly air- or bimodal-breathing fishes. Strictly air-breathing actinopterygian fish are represented in the Amazon by *Arapaima gigas* (Osteoglossiformes) and by *Electrophorus electricus* (Gymnotiformes), both belonging to the

Neopterygii group (see Fig. 4.1). A long and independent evolutionary period separate these three air-breathing fish species. Dipneusti seem to have appeared during the Lower Devonian some 250 Ma ago, while the Neopterygii had its main radiation event during the Mesozoic, and Osteoglossiformes is much older than Gymnotiformes.

Interestingly, they have developed three different physical adaptations which allow for the uptake of O_2 directly from the atmosphere. A true lung

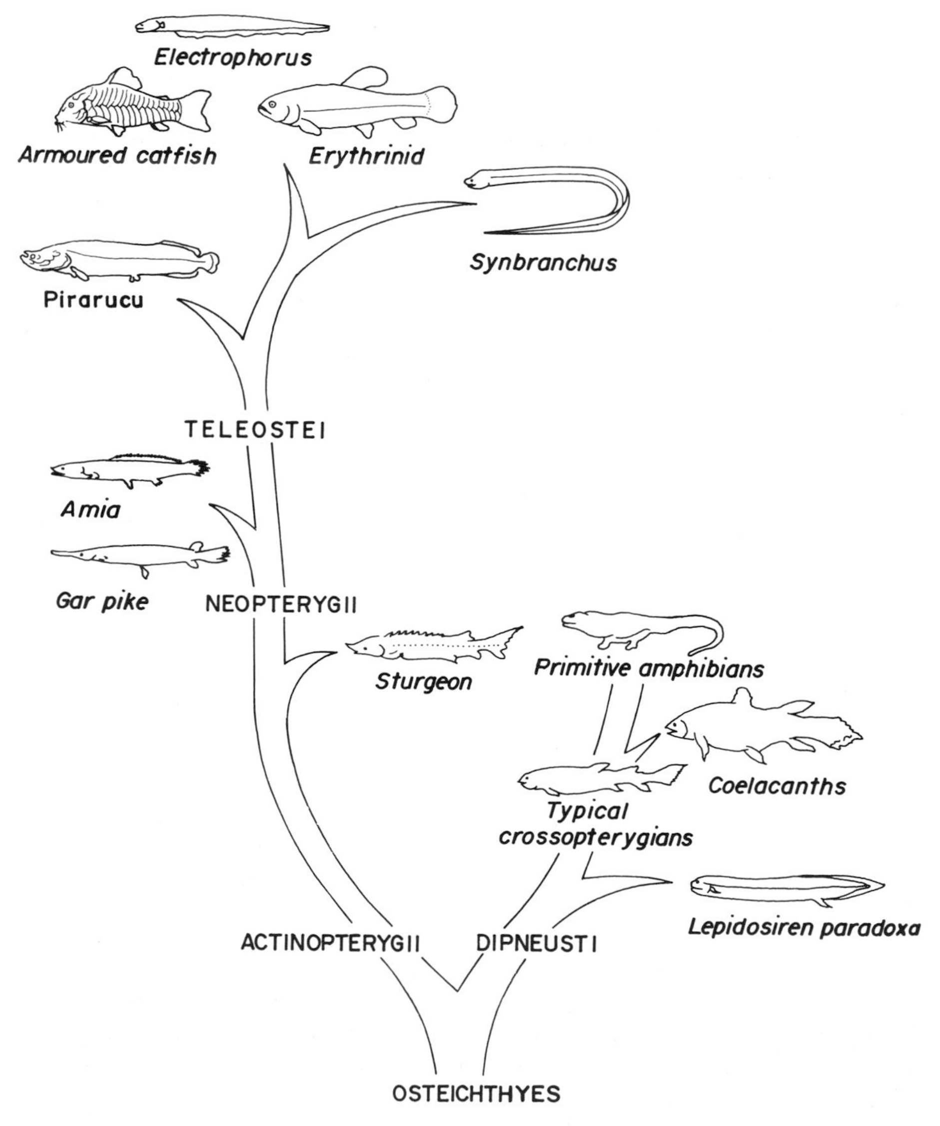

Fig. 4.1. View of the Osteichthyes showing the main groups of air-breathing fishes of the Amazon. The hierarchy is according to Nelson (1984)

is present in *Lepidosiren*, while a modified swim bladder and extensive mouth vascularization are present in *Arapaima* and *Electrophorus*, respectively. The air-breathing organs in all three species, however, appear to be used mainly for O_2 uptake. Gills and skin are used for CO_2 excretion and for ion, water, and pH regulation (Johansen et al. 1978a; Randall et al. 1978a, 1981).

4.1.1 The Lungfish *Lepidosiren paradoxa*

The air-breathing capacity of *Lepidosiren* and *Protopterus* is at a much more advanced stage than the Australian *Neoceratodus forsteri* in which the gills are still responsible for the major part of gas exchange (Johansen and Lenfant 1967; Burggren and Johansen 1987). *Lepidosiren* is primarily an air-breather; it can endure months of almost exclusively terrestrial life during its estivation period (Carter and Beadle 1931c; Sawaya 1946). If it is completely deprived of water, however, the skin quickly dries and the animal survives only for a few hours (Carter and Beadle 1931c; pers. observ.).

During the estivation period, which coincides with the lower water level period in the Amazon basin, *Lepidosiren* digs a burrow about 1 m deep. The opening of the burrow is commonly covered with mud and the estivation chamber in the bottom of the burrow is large enough to accommodate the animal. The burrow is sometimes blocked at regular intervals; only small holes are left open, probably to allow for O_2 diffusion (Carter and Beadle 1931c). These South American lungfish normally estivate in wet substrata and therefore do not form a cocoon, as do the *Protopterus* whose cocoon is formed by a skin secretion. It seems to be essential to keep the animal wet in order to protect it against microbial infection from the soil (Burggeen et al. 1985).

As the water level begins to rise, *Lepidosiren* leave the burrows and typically make a nest with macrophytes, leaves, and twig pieces. The eggs are laid in the nest which is often very hypoxic. Males normally stay with the eggs and stir the surrounding water in order to increase the amount of O_2 available to the eggs. At this time, the males develop a considerable number of respiratory filaments on the pelvic fins and therefore, do not have to surface (see Phelps et al. 1979a). According to Carter and Beadle (1931c), two main changes occur in the larvae after the eggs hatch. The first change is the degeneration of external gills which occurs about 45 days after hatching and has no noticeable effect on growth rate of the larvae. The second change occurs when the larvae start to take in food through their mouths, about 15 days after the degeneration of the gills. At this time, there is a small decrease in their growth rate. Based on these observations, Carter and Beadle (1931c) concluded that the tissue oxygenation is efficiently maintained by the gills during the yolk absorption process.

According to Johansen (1970), the type of lung found in *Lepidosiren* and *Protopterus* resembles that present in lower vertebrates. The air space is divided by a series of septa, ridges, and pillars into smaller compartments which are further subdivided into alveoli-like pockets. These pockets are covered with blood vessels. *Lepidosiren* have two lungs which are fused anteriorly forming a common pneumatic duct which opens into the pharynx at the glottis. Smooth muscle surrounds the pneumatic duct. In addition, these lungs exhibit many features which are typical of higher vertebrates, for example they have pulmonary mechanoreceptors (DeLaney et al. 1983). *Lepidosiren* normally fill the lungs with a single air-breathing sequence every 4–5 min. Bishop and Foxon (1968) however, reported multiple buccal pumping for this species. The O_2 tension in the lungs drops almost linearly during the first 5 min after a spontaneous breath, i.e., it drops from about 150 to less than 50 mm Hg. Interestingly, the O_2 tension of the pulmonary blood remains virtually unchanged, remaining at about 50 mm Hg during this period. However, it drops significantly if the animal does not surface to fill the lungs again (Johansen and Lenfant 1967; pers. observ.).

Based on anatomical evidence, it has been suggested that the partitioning of the heart into left and right parts is most complete in *Lepidosiren* (see Johansen and Lenfant 1967; Burggren and Johansen 1987). In addition, Foxon (1950) showed there is almost no interruption between ventral and dorsal sides of arterial arches which are seen as a bypass of the branchial circulation. It is possible that this bypass is completed during yolk absorption and contributes to the gill's degeneration. Transverse sections of the reduced gill filaments showed that they are poorly vascularized (Fullarton 1931). Based on the O_2 tension of the dorsal aorta and pulmonary artery in specimens exposed to normoxia, hyperoxia, and inhaling pure O_2, Johansen and Lenfant (1967) clearly confirmed the selective passage of blood through the heart. These authors suggested that in *Lepidosiren* the pulmonary blood is directed to the anterior branchial arteries and then to the systemic circulation while the systemic blood is directed to the posterior branchial arteries and then to the pulmonary arteries (see Fig. 4.2). These three characteristics, i.e., the degeneration of the gill, the bypass between ventral and dorsal sides of arterial arches, and the selective passage of the blood through the heart reduce the amount of O_2 lost across the gills. In addition, O_2 loss through the respiratory filaments on the pelvic fins of the males seems to be reduced by ceasing aerial respiration and thus decreasing the O_2 tension of the blood.

As expected, this design results in a significantly higher concentration of blood CO_2 in *Lepidosiren* than in water-breathing fishes living in well-aerated water where the fish's CO_2 is removed mainly through the gills and the skin (see Johansen 1970; Randall et al. 1981). Lenfant et al. (1970) estimated that air-breathing accounts for about 30% of CO_2 removal in *Lepidosiren*. In addition, the branchial respiratory movements are almost imperceptible in this species when they are at rest (Johansen and Lenfant

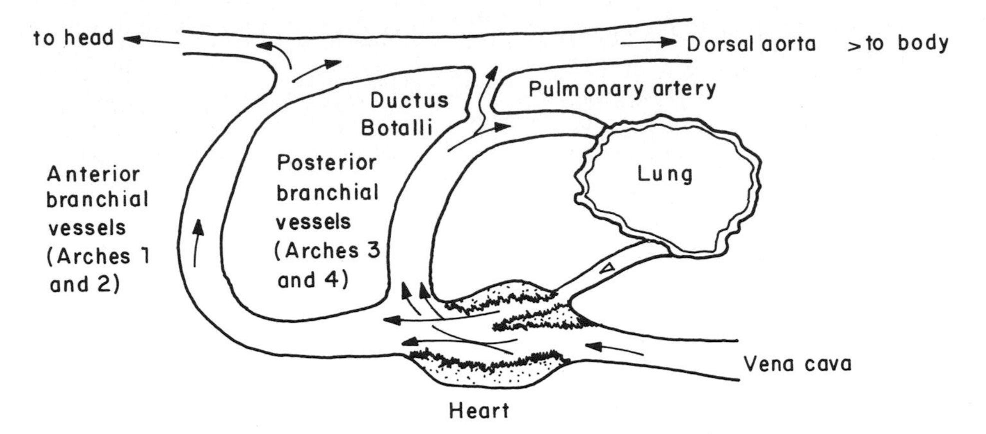

Fig. 4.2. Schematic diagram of the anterior circulation of the lungfish. (Redrawn from Randall et al. 1981)

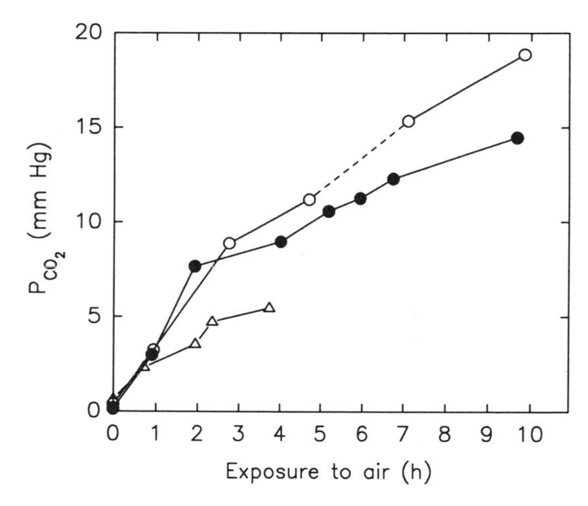

Fig. 4.3. Time-dependent increase in CO_2 tension for specimens of the lungfish of the Amazon, *Lepidosiren paradoxa*, exposed to air in a closed respiration chamber. (Redrawn from Johansen and Lenfant 1967)

1967); this further reduces the CO_2 excretion. When the animals are exposed to air and the gills become nonfunctional, a time-dependent increase in P_{CO_2} in a closed respiration chamber is observed (Fig. 4.3). According to Randall et al. (1981), this pattern is important in the control of HCO_3^- movements and therefore, body pH. It is unknown, however, how the animals deal with CO_2 removal during the estivation period. During this period they keep breathing air but are exposed to a small amount of water which may become highly saturated with CO_2. It is possible however, that the water in the estivation chamber is slowly replaced due to changes in underground water levels.

As previously mentioned, air-breathing accounts for almost 100% of O_2 uptake in *Lepidosiren*. Low hemoglobin-oxygen (Hb–O_2) affinity has been observed for many air-breathers but, *Lepidosiren*, in contrast, seems to operate with a relatively higher Hb–O_2 affinity. In fact, the P_{50} value at pH 7.4 is about 8 torr for *Lepidosiren*, while it is around 21 for *Arapaima*, and 25 torr (pH 7.6) for *Protopterus* (Johansen 1970; Johansen et al. 1978b; Phelps et al. 1979a). Even in the presence of 5.6% CO_2, when the pH of the sample drops to 7.33 and P_{50} increases up to 14 torr (Phelps et al. 1979a), the Hb–O_2 affinity in *Lepidosiren* is still higher than that observed for air-breathers (see Table 4.2). No Root effect (see Sect. 4.1.4) has been reported for the South American lungfish.

Studies of whole blood have indicated a decrease in the magnitude of the Bohr factor in this species. Analyzing juveniles of *Lepidosiren*, Johansen and Lenfant (1967) reported a Bohr factor ($\phi = \Delta P_{50}/\Delta pH$) of -0.234, while Lenfant et al. (1970) reported $\phi = -0.295$ for adults, which is in the same range ($\phi = -0.30$) as that detected by Phelps et al. (1979a) during the Research Vessel Alpha Helix expedition to the Amazon. With the exception of *Protopterus*, higher magnitudes of the Bohr factor have been observed in other air-breathers (Table 4.2). The Bohr factor, however, is highly influenced by ATP concentrations in *Lepidosiren*; a fivefold increase in the magnitude of the Bohr factor in the presence of 1 mM of ATP was reported by Phelps et al. (1979a). ATP also induces a significant increase in the cooperativity values (n) between pH 7 and 8. This clearly indicates that *Lepidosiren* is able to regulate the Hb–O_2 affinity by changing the intraerythrocytic concentration of ATP. However, to what extent the ATP levels are regulated during estivation or even in the males during the breeding period remain to be investigated.

The erythrocytes of *Lepidosiren* also contain a significant amount of GTP (Fig. 4.4). The GTP:ATP ratio can be as high as 0.78 (Bartlett 1978a). The distinctive feature, however, is the presence of inositol diphosphate (IP_2) which is also present in *Protopterus* (Bartlett 1978a; Isaacks et al. 1978). Its intraerythrocytic concentration was about 50% of that of GTP, i.e., 8% of all phosphates for the analyzed animals. It should be noted that changes in the concentration of red blood cell organic phosphates are very common and have been related to many physiological and environmental factors (see below). Red blood cells of lungfish, both South American and African, also contain unusually large amounts of uridine nucleotide, mostly UTP and UDP. Additionally, a large pool of fructose-diphosphate was observed in one specimen of *Lepidosiren* (Bartlett 1978a). The physiological significance of these compounds in these air-breathers is yet to be evaluated.

In summary, even though *Lepidosiren* is almost independent of water as a source of O_2, it has preserved many of the characteristics typical of water-breathers which include CO_2 removal through the gill and skin, high Hb–O_2

Table 4.2. Magnitude of the Bohr factor and O_2 affinities of hemolysates of selected obligatory and facultative air-breathing fishes of the Amazon

Species	Hb (g%)	Ht (%)	O_2 content (vol%)	P_{50} at pH[a]	Bohr effect	References
Lepidosiren (juvenile)	5.4[a]	19	6.80	10.8 (7.6)	−0.234	Johansen and Lenfant (1967)
Lepidosiren (adult)	6.6	28	8.25	–	−0.295	Lenfant et al. (1970)
Lepidosiren (stripped Hb)	–	–	–	1.27 (7.0)	−0.180	Phelps et al. (1979a)
Arapaima	–	35	–	16.33 (7.4)	−0.390	Galdames-Portus et al. (1979)
Arapaima	–	–	–	21 (7.4)	−0.300	Johansen et al. (1978a)
Hypostomus (normoxia)	10.4	31.5	11.15	18.5 (7.6)	−0.320	Weber et al. (1979)
Pterygoplichthys	10.9	36	13.6[a]	15.4 (7.5)	−0.340	Val et al. (1990); ALV (unpubl. data)
Electrophorus	11.2	41	13.90	10.7 (7.57)	−0.680	Johansen et al. (1968)
Synbranchus	13.0	47	17.30	7.2 (7.6)	−0.400	Lenfant et al. (1970)

[a] Calculated from O_2 capacities (1.25 ml O_2/g Hb).

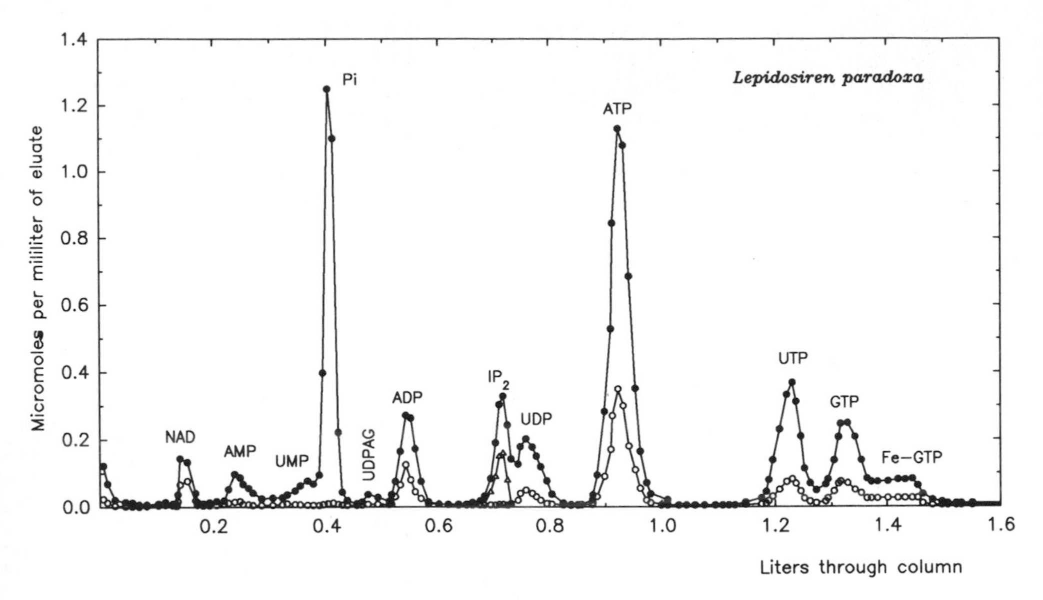

Fig. 4.4. Ion chromatographic profile of intraerythrocytic phosphates of the lungfish of the Amazon, *Lepidosiren paradoxa*. Observe the significant amount of IP_2 and UTP. Quantities are expressed in μM/ml of eluate. ● Total phosphorus; ○ optical density at 260 nm calculated as micromoles of adenine; △ inositol color method. (Redrawn from Bartlett 1978a)

affinity, and the presence of ATP and GTP as strong modulators of $Hb-O_2$ affinity. These characteristics seem to be essential during the estivation and breeding season. Information about either the presence or absence of carbonic anhydrase in the lung cells, regulation of ATP and GTP levels during very specific phases of their life, breathing control, and the effect of circulating catecholamines on respiratory parameters (if they are released) is essential for a more complete evaluation of gas exchange processes in *Lepidosiren*.

4.1.2 *Arapaima gigas*

In contrast to *Lepidosiren*, *Arapaima gigas* uses a highly modified swim bladder to take in O_2 from the air. The modified swim bladder is similar to the lungs observed in the true lungfishes (Fig. 4.5), however, only one chamber is present in the swim bladder. The inner wall is subdivided by septa into small chambers whose walls are highly vascularized. The large physostome swim bladder is positioned dorsally above the abdominal cavity, next to the kidney which is separated from the rest of the body by peritoneum (Migdalski 1957; Hochachka et al. 1978a; Johansen 1979). Interestingly, the lipid content, as a percent of dry weight, is reduced in *Arapaima* and

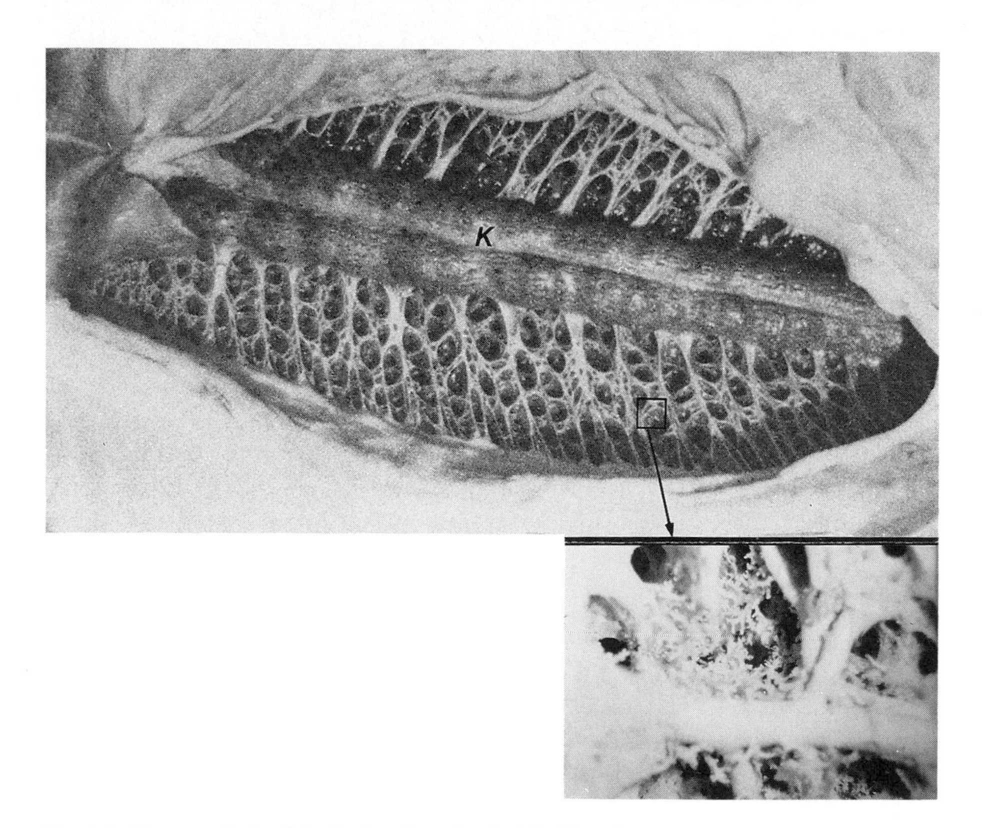

Fig. 4.5. The vascularized single-chambered swim bladder of *Arapaima gigas*. Note the cellular structure and the well-developed kidneys (*K*). The *inset* shows the division of the inner wall into small compartments. (Modified from Hochachka et al. 1978a; Randall, unpubl.)

Lepidosiren compared with other aquatic vertebrates (Phleger and Saunders 1978), both having large air spaces which will increase buoyancy.

Blood supply to the swim bladder of *Arapaima* is via a series of arteries which regularly branch off along the length of the dorsal aorta (Farrell 1978). The oxygenated blood returning from the swim bladder enters the heart along with the blood returning from the systemic circuit. The mixed blood is then pumped through the gills before entering the systemic circulation and the swim bladder (Fig. 4.6).

Arapaima surface every 4–5 min for a breath (Farrell and Randall 1978). To surface, the animal normally lifts its snout, always keeping its head higher than its tail, and then moves its head out of water to ventilate. The typical sound that is emitted at this time is often recognizable by local fishermen to locate and harpoon the animal. According to Farrell and Randall (1978), when the animal reaches the water's surface a small pressure gradient favours exhalation. The buccal floor is then lowered and the mouth is opened, drawing air into the mouth. When the pneumatic duct is opened

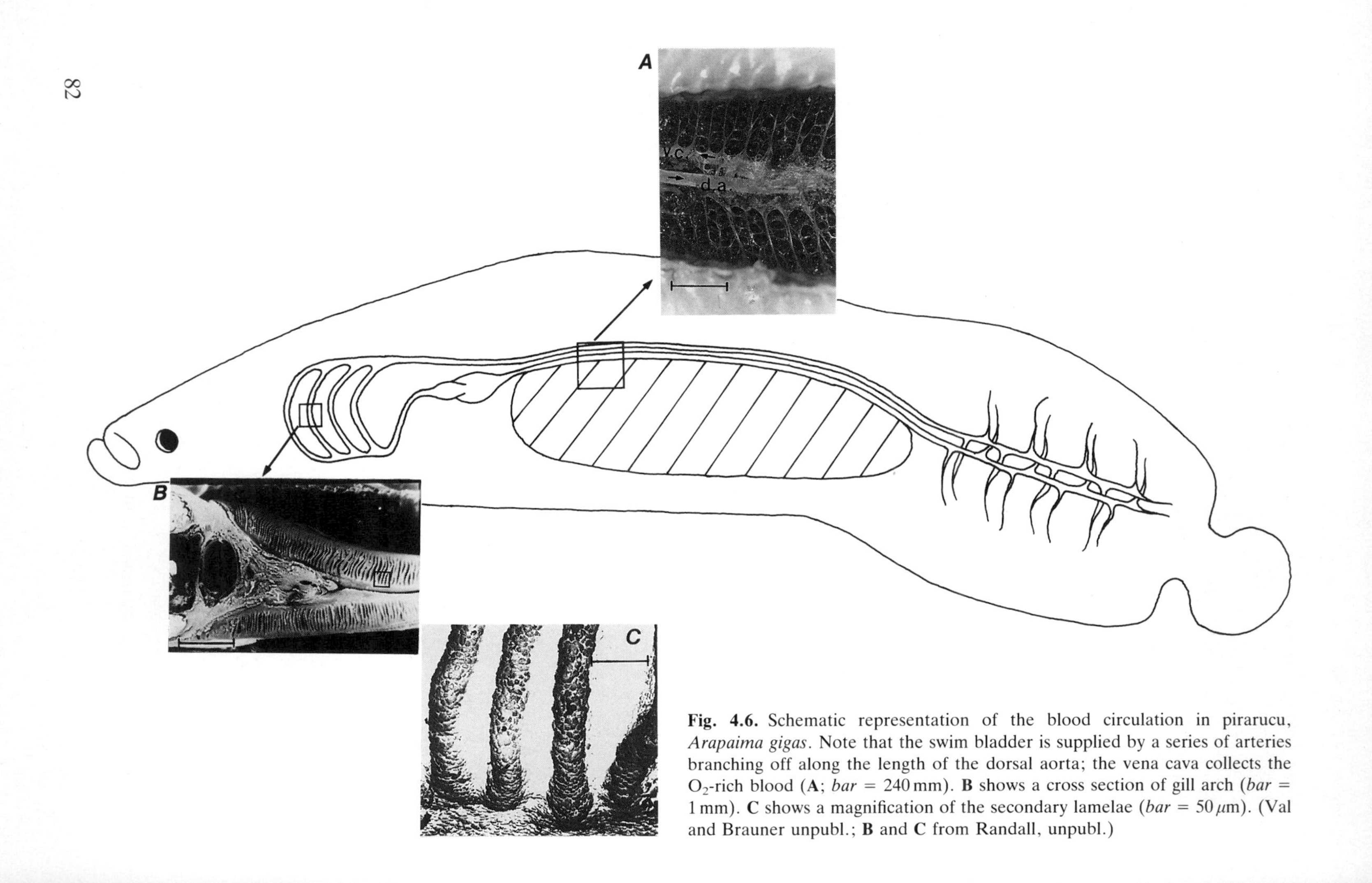

Fig. 4.6. Schematic representation of the blood circulation in pirarucu, *Arapaima gigas*. Note that the swim bladder is supplied by a series of arteries branching off along the length of the dorsal aorta; the vena cava collects the O_2-rich blood (**A**; *bar* = 240 mm). **B** shows a cross section of gill arch (*bar* = 1 mm). **C** shows a magnification of the secondary lamelae (*bar* = 50 μm). (Val and Brauner unpubl.; **B** and **C** from Randall, unpubl.)

almost simultaneously, air is aspirated into the air bladder. It is thought that exhalation is passive and inhalation is the result of bladder aspiration combined with the action of a buccal pump. The aspiration is generated by a diaphragm-like membrane stretching between the body flanks of the animal. As the animal curves its head at the water's surface, a dorsal–ventral pressure on its vertebral column induces an outward movement of the flanks which pulls the membrane downwards generating subatmospheric pressures in the swim bladder (see Farrell and Randall 1978 for further details).

The gills of *Arapaima* are small in size (Fig. 4.6) even compared with its close relative aruanã (*Osteoglossum bicirrhosum*). According to Hulbert et al. (1978a), some adjustments of the gill structure controls blood flow in *Arapaima*. These adjustments include thick layers of smooth muscle on the walls of efferent and afferent arteries; a branched secondary arteriole which emanates from the afferent branchial artery and perfuses a secondary lamella via several discrete lacunar openings composed of smooth muscle; and the position of this secondary arteriole. These features are designed to short-circuit the gill circulation in order to help in reducing the O_2 loss via gill circulation. It is thought that the heart circulation in *Arapaima* does not provide the same degree of blood separation as in *Lepidosiren*.

Another striking factor described by Hulbert and coworkers in the gills of both *Arapaima* and aruanã is the presence of an inactive-type of chloride cells on the primary gill epithelium. These chloride cells, however, exhibit different effects on ion-regulated ATPases in the two species. For example, the total Na^+/K^+ ATPase activity in aruanã is 12 times greater than in *Arapaima*. Thus, the capacity of the gills in *Arapaima* for ion regulation seems to be reduced, too. Based on the characteristics of the water (low pH, low dissolved ion levels, etc.), it is possible to anticipate a low activity of gill proton ATPase. Proton ATPase seems to generate a potential to drive the sodium influx from the water and seems to be influenced by gill pH changes (see Lin and Randall 1990; Randall and Val 1993). Ion regulation, however, seems to be carried out mainly in the kidney in *Arapaima* (Hochachka et al. 1978a) probably as an adaptation to the water characteristics.

About 78% of the O_2 is taken up from the air via the swim bladder in *Arapaima* when exposed to normoxia. This fraction increases to 100% when the animal is exposed to anoxic water (Stevens and Holeton 1978a). However, when exposed to an atmosphere with varying levels of O_2, the animal is unable to increase its O_2 uptake from normoxic water, via gills or skin, beyond 25%. In addition, the animal will die in a few minutes when exposed to a pure nitrogen atmosphere or if denied access to air (ALV, unpubl. data). These observations clearly indicate that the O_2 uptake from water (even at maximum levels) does not support the metabolic needs in *Arapaima*. However, they do suggest that the control of gill circulation, as described above, does not contribute to an increase in O_2 uptake from water. Thus, *Arapaima* relies on the O_2 uptake from air via the swim bladder to survive.

The gills, on the other hand, seem to be an important site for CO_2 excretion. According to Randall et al. (1978a), about 63% of the total CO_2 is excreted into the water via gills and only 37% via the swim bladder (Fig. 4.7). These authors suggested that the $CO_2:HCO_3^-$ system in the blood (from swim bladder and from systemic circulation) which is pumped to the gills is in equilibrium. Because a significant part (63%) of the total CO_2 is excreted through the gills, the blood in the dorsal aorta is not in a state of equilibrium. Part of this blood enters the swim bladder circulation where the animal gets rid of the rest (37%) of the CO_2 which needs to be removed. It is possible that this proportion (63:37, G:S) is much more related to the circulation design than to the efficiency of the gills in the removal of CO_2. In other words, if the gills become completely inactive, it may be possible that almost 100% of the CO_2 will be removed at the swim bladder. If this is true, the gills in *Arapaima*, which are already reduced in size and play a minor role in ion regulation, are not absolutely necessary for CO_2 excretion.

The P_{O_2} in the swim bladder increases to about 140 mm Hg just after the aspiration and nearly all gas exchange occurs during the first minute following a breath (Fig. 4.8). After 5 min the P_{O_2} in the swim bladder is still about 50 mm Hg. At this P_{O_2} level the Hb system in *Arapaima* is nearly 90% saturated (Johansen et al. 1978b). Since the P_{50} of red cell suspension is about 21 mm Hg (Fig. 4.9) the animal can delay aspiration if the surfacing cost is high (see Kramer 1988 for further details). In fact, although the animal surfaces spontaneously every 4 min, it can survive up to 10–12 min if denied access to water surface (pers. observ.). The time between aspirations is limited because the arterial blood would be only about 25% saturated

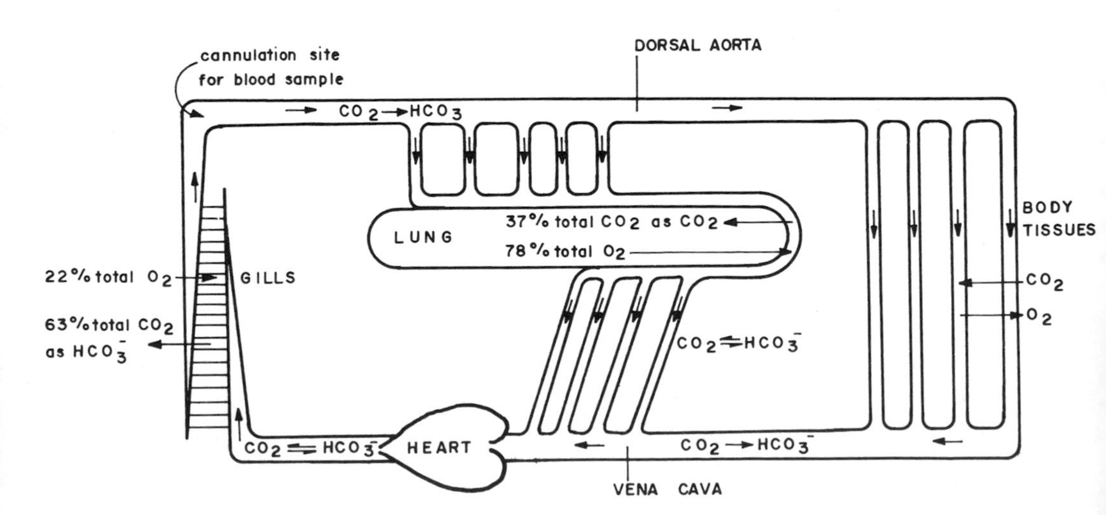

Fig. 4.7. Schematic representation of gas transfer pathways in pirarucu, *Arapaima gigas*. Note that the O_2 is primarily taken up in the swim bladder while the CO_2 is excreted through the gills as bicarbonate (HCO_3^-). (Redrawn from Randall et al. 1978a)

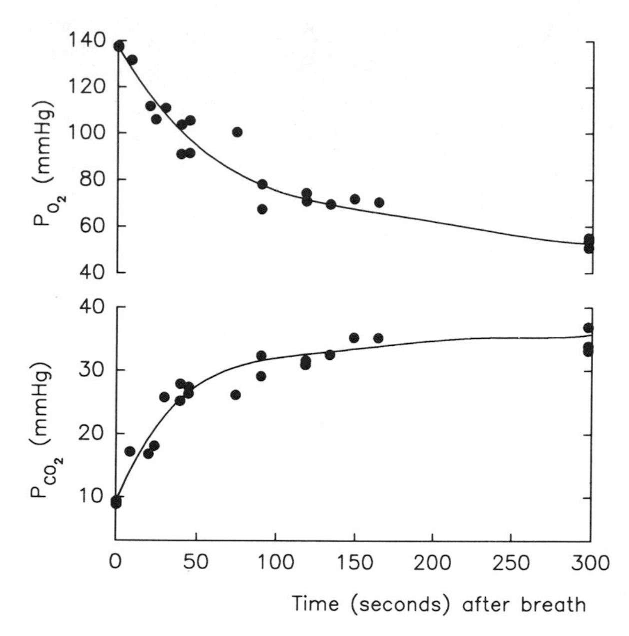

Fig. 4.8. Time-course changes in P_{O_2} and P_{CO_2} in the swim bladder os pirarucu, *Arapaima gigas*, after an air breath. (Redrawn from Randall et al. 1978a)

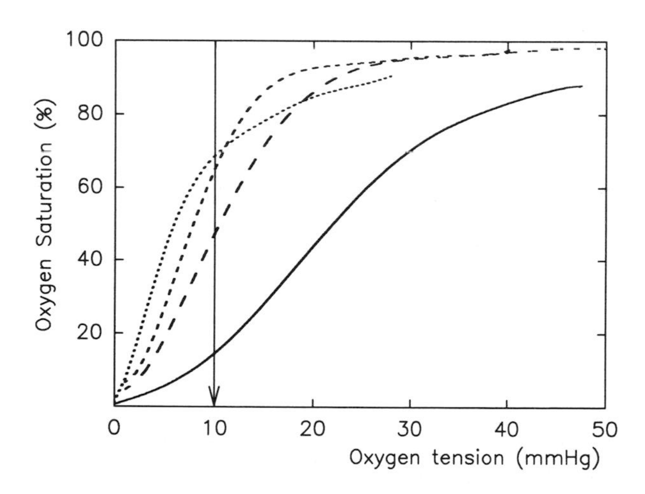

Fig. 4.9. Oxygen equilibrium curves of red cell suspensions for selected air-breathing fishes of the Amazon. Note the lower P_{50} values for *Arapaima gigas* (*solid line*). *Osteoglossum bicirrhosum* (*short-dashed line*) is a water-breather and has been included for comparison. *Medium-dashed line*, *Hoplosternum littorale*; *long-dashed line*, *Synbranchus marmoratus*. The *arrow* indicates the lowest diurnal O_2 tension observed in the várzea of the Marchantaria Island, Solimões river, during 1989. (Data compiled from Johansen et al. 1978b; Phelps et al. 1979b; Garlick et al. 1979)

(minimum) if the animal were to rely exclusively on dissolved O_2. Typically, the O_2 levels in their natural habitat ranges between 15 and 30 mm Hg (see Johansen et al. 1978b for further details).

One major and two minor Hb fractions have been detected in all of the specimens of *Arapaima* that we analyzed using starch gel electrophoresis. Fyhn et al. (1979), however, reported two phenotypes (I and II) using polyacrylamide as the support. These hemoglobins exhibit higher P_{50} values than the typical water-breathers. Compared with aruanã, for example, the P_{50} value for stripped Hb of *Arapaima* is about seven times higher (Johansen et al. 1978b), reflecting the values observed for red cell suspensions (see Fig. 4.9). The magnitude of the Bohr factor ($\phi = -0.30$) is in the same range of that detected for *Lepidosiren*. A significant Root effect, about 47%, is observed (see Sect. 4.1.4) when the blood pH is decreased from 7.8 to 5.5 (Farmer et al. 1979; Galdames-Portus et al. 1979; ALV, unpubl. data).

In addition to the pH sensitivity, the Hb system of *Arapaima* is highly affected by the main organic phosphates. Johansen et al. (1978b) showed that GTP is a more effective cofactor than ATP in all Hb:NTP ratios analyzed. In addition, they showed a significantly greater $Hb-O_2$ depression in the presence of inositol hexaphosphate (IHP), which has six negative charges compared with three in GTP and ATP. This is of particular interest because a significant amount of inositol pentaphosphate (IPP) has been detected in the erythrocytes of this fish species (Isaacks et al. 1977; Bartlett 1978b).

4.1.3 *Electrophorus electricus*

In concurrence with the previously analyzed examples, *Electrophorus* is also an obligatory air-breather. However, this species uses a completely different structure for the uptake of O_2 from the air. The air-breathing organ in *Electrophorus* is the mouth (Fig. 4.10). It is structurally adapted for air-breathing by an extensively diverticulated and richly vascularized mucosa which is papillated and distributed over both the floor and the roof of the mouth. In the floor of the mouth there are three rows of papillae while in the roof there are four rows. They are arranged in a such way that when the mouth is closed they fit into each other. Some smaller prominences are present also on the branchial arches and on the lateral branchial walls (Johansen et al. 1968). Based on formalin-fixed material, these authors estimated the surface area of the respiratory mucosa to be about 15% of the total body surface, this is a smaller surface area than that found in either the gills of specialized water-breathers or in the lungs of vertebrates.

Similar to other air-breathers, the gills of *Electrophorus* exhibit reduced anatomical development. The primary filaments are exceptionally reduced in size, the secondary filaments are narrow and thick, and the epithelial tissue is somewhat coarse. As a consequence, they play only a minor role in

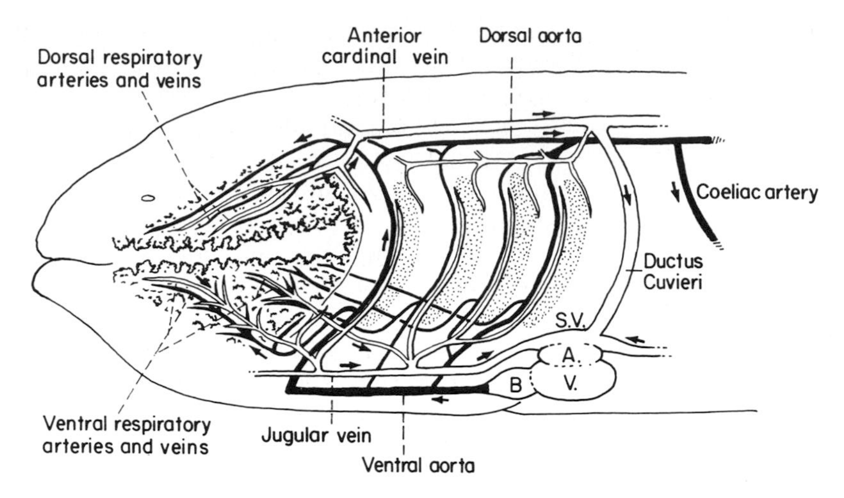

Fig. 4.10. Schematic representation of the vascularized and papillated mucosa of the gymnotid *Electrophorus electricus*. *A*. Atrium; *B*. bulbus arteriosus; *V*. ventricle. (Redrawn from Johansen et al. 1968)

gas exchange. In addition, an extensive shunt is present in the gills further reducing their gas exchange ability (Johansen et al. 1968; Johansen 1982). Thus, it is not surprising that the animal will die in about 20 min if denied access to air.

The location of a delicate structure such as respiratory epithelium in the mouth is a disadvantage due to its susceptibility to mechanical injury (Johansen et al. 1968). Nevertheless, it accounts for 78% of the total O_2 needed for the animal (Farber and Rahn 1970). The animal surfaces every 2 min to take a breath which consists of a quick exhalation and inhalation of air through the mouth opening followed by a quiet period of submergence (Garey and Rahn 1970). In accordance with the findings of Garey and Rahn (1970) and Johansen et al. (1968), we did not observe gas displacement through the opercular opening when the fish were below the water's surface. It is important to point out, that, in contrast to other air-breathers, opercular movements in *Electrophorus* were not observed. According to Garey and Rahn (1970), the rest of the necessary O_2 (22%) is taken up through the skin. However, no noticeable effect on respiratory efforts is observed when the animal is exposed to hypoxic water. When exposed to hypoxic atmosphere, on the other hand, a significant increase in air-breathing frequency is observed (Johansen et al. 1968).

The mouth and skin are also major sites for CO_2 excretion in *Electrophorus*. The gills play only a small role in CO_2 excretion. According to Farber and Rahn (1970), about 81% of the CO_2 is excreted through the skin, while 19% is excreted through the mouth resulting in a high gas exchange ratio (3.7) for the skin compared to that of the mouth (0.25). These values are respectively 2.11 and 0.35 for *Arapaima* (Randall et al.

1978a) and 8.80 and 0.40 for *Lepidosiren* (Lenfant et al. 1970). So, in all three obligatory air-breathing fish of the Amazon, the air-breathing organ is the major site for O_2 uptake but not for CO_2 removal. The same general trend, has been extensively observed in other bimodal air-breathers (Rahn and Howell 1976).

The importance of the skin as the major site for CO_2 excretion in *Electrophorus* was first suggested by Johansen et al. (1968) based on results showing insignificant increases in blood P_{CO_2} in animals inspiring air containing elevated levels of CO_2. The importance of water convection over the skin for the elimination of CO_2 was shown later by Garey and Rahn (1970) who based their conclusions on blood P_{CO_2} changes in animals exposed to air and water. No opercular movements but extensive water flow over the skin due to movement have been observed in this scaleless fish species.

In spite of the efficiency with which the skin removes CO_2, blood P_{CO_2} and plasma HCO_3^- levels are extremely high. Garey and Rahn (1970) reported P_{CO_2} of 29 mm Hg and plasma HCO_3^- of 30.2 mM/l. Johansen et al. (1968) reported similar P_{CO_2} levels for arterial blood. These values are even higher than those reported for other air-breathers. Arterial blood pH (7.58, Garey and Rahn 1970), however, is similar to that reported for other fish species at the same temperature. As a consequence, a significantly high buffering capacity is observed in *Electrophorus*. In addition, a significantly higher Hb level has been observed in this fish species, possibly indicating an additional buffering capacity. Considering the sluggish life-style and the characteristics of its habitat (low O_2 and very little or no water movement), this extremely high buffering capacity seems to be very important for maintaining blood pH because the fish body surface is washed only when the animal moves itself to the water's surface to breathe.

Hematocrit and O_2 capacity are also augmented in this sluggish fish species (Johansen et al. 1968). These authors suggest that these characteristics plus the high Hb–O_2 affinity are related to the enormous shunting of oxygenated blood to the systemic veins. In addition, the magnitude of the Bohr effect ($\phi = -0.680$) is higher than the normal observed for water-breathing fish or even for the Amazonian obligatory air-breathers which were previously described in this chapter. No Root effect (see Sect. 4.1.4) has been reported for the Amazonian electric eel.

A high Bohr factor is an important adaptive measure against a high buffer capacity such as that observed in *Electrophorus*. It means that the Hb is able to unload the O_2 to the tissues even when the pH decreases slightly. This combination of high Bohr factor and high buffering capacity is seen in *Electrophorus*. In addition, it is possible that the intraerythrocytic phosphates have the same effect on Hb–O_2 affinity in the electric eel as they have in other fish species of this group. If this is true, in the presence of a modulator, a higher Bohr factor could be expected to occur in order to further facilitate O_2 delivery to the tissues.

4.1.4 Controversial Points

A low or null Root effect seems to be a general characteristic of air-breathers, except for *Arapaima gigas* (Table 4.3). According to Root (1931), the Root effect is simply the reduction in the blood's ability to carry O_2 when it has been exposed to acidic conditions. Indeed, some microenvironmental conditions (presence of allosteric effectors, for example) can stress the Root effect of some hemoglobins (ALV, unpubl. data; Farmer et al. 1979). It was originally assumed that the physiological role of the Root effect is facilitate O_2 release into the swim bladder. In the obligatory air-breathers, however, neither is the swim bladder filled up with O_2 released from the blood nor is the pH significantly reduced. It is quite possible that the air-breathing habit evolved simultaneously with the development of smaller effects of pH on the O_2-carrying capacity, mainly because the high buffer capacity observed in these animals. The physiological rationale for a high Root effect in *Arapaima* (Galdames-Portus et al. 1979), however, is still unknown. On the other hand, obligatory air-breathing fishes exhibit a large Bohr effect which is advantageous considering the observed high buffer capacity.

Another interesting aspect of obligatory air-breathing fishes is that their system for O_2–CO_2 exchange is neither coupled in the timing of exchange nor in the site of exchange. In general, as the blood flows through the gills of water-breathing fish the oxygenation of Hb causes the release of protons which drive the intraerythrocytic dehydration of bicarbonate; the CO_2 thus formed is excreted. This reaction is catalyzed by the erythrocytic enzyme carbonic anhydrase. Thus, there is a linkage between O_2 uptake and blood

Table 4.3. Amplitude of the Root effect of selected fish species of the Amazon. (Data from Val and Peixoto-Neto, unpubl.)

Families and species	Amplitude Root effect	Breathing mode
Lepidosireniformes		
Lepidoiren paradoxa[a]	0	Air
Osteoglossiformes		
Arapaima gigas	47	Air
Siluriformes		
Hoplosternum littorale	10	Facultative
Pterygoplichthys sp.	15	Facultative
Gymnotiformes		
Electrophorus electricus[a]	0	Facultative
Synbranchiformes		
Synbranchus marmoratus[a]	0	Facultative

[a] Data from Farmer et al. (1979).

CO_2 removal (Perry and Laurent 1990; Randall 1990; Randall and Val, in press). In the obligatory air-breathing fishes however, most of the O_2 is taken up in the air-breathing organ while the bulk of CO_2 is excreted through the gills or skin. Consequently, CO_2 removal does not occur simultaneously with O_2 uptake in these animals. Although the system is uncoupled, there is no evidence that the dehydration of bicarbonate is not driven by the protons released during the Hb oxygenation. If the system is coupled in this way, it is likely that the protons released during the oxygenation of Hb in the air-breathing organ, will then sustain the dehydration of bicarbonate thus creating a high blood P_{CO_2} which will enhance its removal at the gills and/or skin.

4.2 The Facultative Air-Breathers

Facultative air-breathers extract O_2 from air when water O_2 is low. Facultative air-breathing neopterygian fish occur in many different Amazon families. This group of fish predominantly use their gills for breathing normoxic water. However, when the waters become hypoxic, they are able to breath air using an accessory air-breathing organ. The development of the air-breathing organ in this group did not always involve the parallel development of an "air sac" or swim bladder. The air-breathing organ is, in general, richly supplied with blood vessels and fine capillaries. In contrast to the obligatory air-breathers, the gills in facultative air-breathers are functional in gas transfer in water, displaying the same general specializations observed for the gills of unimodal water-breathers. Many structures have evolved as accessory breathing organs, including gills, modified stomach and intestine, and vascularized swim bladder. Table 4.4 lists some facultative air-breathing fishes of the Amazon, type of accessory air-breathing organ, and some information about their habitats and life-style.

4.2.1 Gills as Air-Breathing Organs

In general, gills are primarily designed to function only in water. When the gills are exposed to air the delicate, fine filaments collapse and the whole structure become nonfunctional for gas exchange. This is irreversible for almost all water-breathing fishes. However, a few species, such as *Synbranchus marmoratus*, the swamp or mud eel, are able to use their gills as an air-breathing organ. Besides *Synbranchus*, this adaptation occurs in only three or four other fish species including *Hypopomus brevirostris* of the Paraguayan Chaco (Carter and Beadle 1931b) and *Anguilla anguilla* (Berg and Steen 1965).

In the Amazon, *Synbranchus* are often found along the edges of ponds and swamps. Sometimes it is possible to observe specimens resting out of

Table 4.4. Air-breathing threshold, accessory structures, and habitat of selected facultative air-breathing fishes of the Amazon

Families and species	Air-breathing accessory structures	Habitat & life-style	Air-breathing threshold[a] (mg O_2/l water)	References
Loricariidae				
Pterygoplichthys multiradiatus	Stomach and intestine	Várzea; sluggish	2.1	Val et al. (1990); Val (unpubl.)
Hypoptopoma gulrae	Stomach and intestine	Várzea; sluggish	0.5	Soares (1993)
Dekeyseria amazonica	Stomach and intestine	Várzea; sluggish	0.4	Soares (1993)
Callichthyidae				
Hoplosternum littorale	Stomach and intestine	Stagnant streams; várzea; sluggish	4.1	Soares (1993)
Hoplosternum littorale	Stomach and instestine	Stagnant streams; várzea; sluggish	Continuous air-breathing	Val et al. (unpubl.)
Hoplosternum thoracatum	Stomach and intestine	Stagnant streams; várzea; sluggish	Continuous air-breathing	Gee (1976)
Callichthys callichthys	Stomach and intestine	Stagnant streams; várzea; sluggish	Continuous air-breathing	Carter and Beadle (1931c)
Erythrinidae				
Hoplerythrinus unitaeniatus	Swim badder and skin	Swamps and streams; active	4 (81 mm Hg)	Stevens and Holeton (1978b)
Synbranchidae				
Synbranchus marmoratus	Mouth, branchial and pharingeal diverticula	Swamps and rivers; sluggish	1.5–2.5 (30–50 mm Hg)	Bicudo and Johansen (1979)

[a] Dissolved O_2 at which aerial respiration is initiated.

water or even moving overland. Similar to *Lepidosiren*, who were discussed earlier in Section 4.1.1, the swamp eels dig a burrow into the mud, close to the edges and estivate when the water level begins to drop. However, *Synbranchus* do not build cocoons, rather, their burrow is labyrinthic with many underground channels. The wide distribution of these eels across South America has been attributed to its overland movements (Kramer et al. 1978).

It is possible that many different populations of *Synbranchus* exist across South America. Animals with different chromosome numbers (2n = 42, 44, and 46) have been observed in different places across São Paulo state (Foresti et al. 1982). In addition, Nakamoto et al. (1986) have shown that the animals with 2n = 44 and 46 chromosomes have a Hb phenotype formed by three fractions, while the animals with 2n = 42 chromosomes exhibit 12 Hb fractions (Nakamoto et al. 1986). These Hb phenotypes are different from those detected by Wilhelm and Reischl (1981) for the animals found in Southern Brazil and those detected by Fyhn et al. (1979) for the animals of the Amazon. Interestingly, the Hb concentration and the mean corpuscular Hb concentration (MCHC) are greater in the animals with 12-banded Hb patterns than in the other phenotype (Nakamoto et al. 1986).

The gills of *Synbranchus* are morphologically specialized to allow for air-breathing. The breadth/height ratio of the secondary lamella is about five times greater than those in other fishes. This feature is designed to prevent the gills from collapsing when they are exposed to air (Carter and Beadle 1931b; Johansen 1966). In addition, the skin and a well-vascularized buccal and pharyngeal mucosa are likely extra sites for gas exchange (Bicudo and Johansen 1979). Thus, the gills constitute the main structure for both aerial– and water–gas exchange. The animal is able to switch between the two modes of gas exchange, i.e, if well-oxygenated water is available the animal is a typical water-breather; if the water O_2 content drops, the animal takes up O_2 from the air. This ability seems to be an efficient adaptation to the regular environmental fluctuations which are observed in the Amazon floodplain areas.

To breathe air, the animal moves slowly to the water surface, expires through both the mouth and the single ventral gill and then inspires, filling the gill chamber with air. The mouth is then closed. This behaviour traps a significant amount of air in the mouth and gill chamber adding buoyancy to the animal, enabling it to remain at the water's surface. Bicudo and Johansen (1979) showed that the O_2 tension of the gas held in the branchial chamber decreases as the CO_2 content increases because the closed mouth prevents the escape of the gas (Fig. 4.11). while the O_2 tension in the gill chamber decreases, an increase in blood P_{O_2} is observed simultaneously with an increase in P_{CO_2} (Johansen 1966).

Heisler (1982), studying *Synbranchus* from the Amazon, showed that the arterial blood P_{CO_2} in fish adapted for air-breathing is about five times higher than in animals acclimated to water-breathing. As a consequence, a

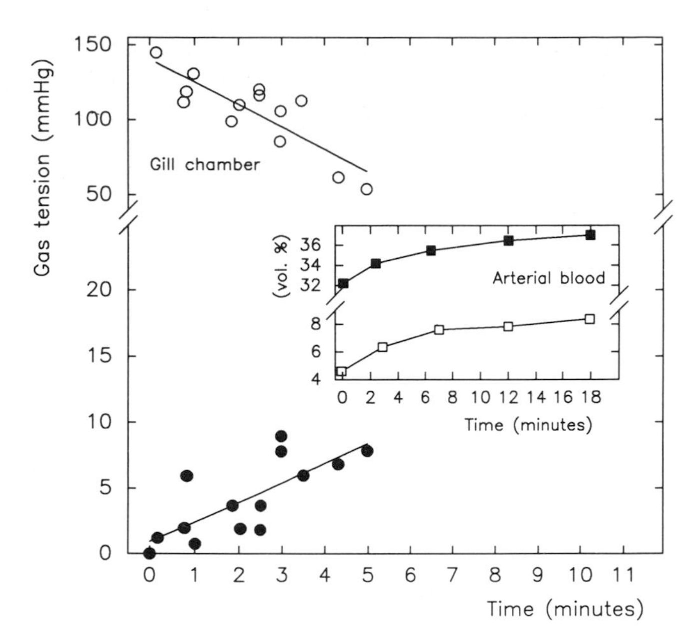

Fig. 4.11. Time course of O_2 (*open symbols*) and CO_2 (*closed symbols*) tensions in the branchial chamber (*circles*) and arterial blood (*squares*) of *Synbranchus marmoratus*. (Redrawn from Johansen 1966; Bicudo and Johansen 1979)

significantly lower plasma pH in air-breathing animals and practically no differences in bicarbonate levels between the two groups were observed. In addition, he showed that when animals who are adapted to water-breathing are exposed to water with a lower content, arterial blood P_{CO_2} increases and plasma pH decreases towards the levels observed for animals adapted to air-breathing. A significant increase in bicarbonate levels is observed during the transition from water- to air-breathing, but they return to the initial levels after about 5 h (Fig. 4.12). According to the author, some animals exhibit a rapid change in these parameters (the group was called "fast") while in others these parameters change slowly (the "slow" group). Heisler (1982) pointed out that these groups of fish ("slow" and "fast") could not be separated by sex, size, or any other apparent parameter. However, different Hb phenotypes have been observed for *Synbranchus* found in the Amazon (Fyhn et al. 1979; ALV unpubl. data).

It is possible that these animals with different phenotypes have different Hb concentrations and mean corpuscular hemoglobin concentration (MCHC), as do the animals from the São Paulo state (Nakamoto et al. 1986). It has been shown that as the severity of the anaemia increases, arterial blood P_{CO_2} increases and plasma pH decreases due to a reduced release of protons for the dehydration of bicarbonate (Wood et al. 1979a; Randall 1982). Although Heisler (1982) was unable to identify two groups

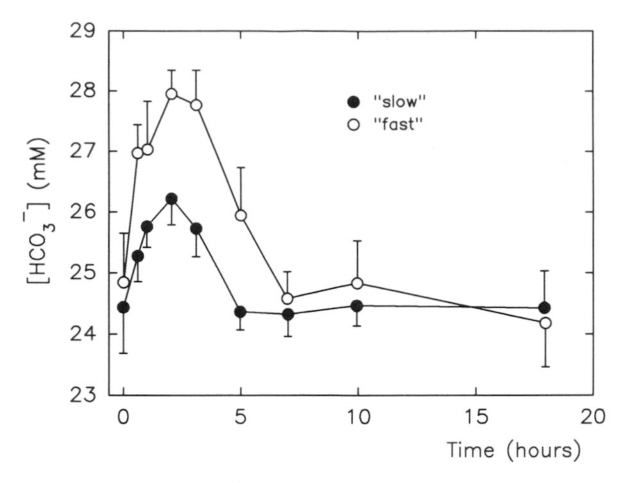

Fig. 4.12. Time course of arterial HCO_3 after initiation of air-breathing in *Synbranchus marmoratus*. (Redrawn from Heisler 1982)

among the adapted animals, it is possible that his sample included animals with different Hb phenotypes and different mean Hb concentrations. If this is true, we suggest that his "fast" animals have a low Hb concentration (phenotype I of Nakamoto et al. 1986), while his "slow" animals have a high Hb concentration (phenotype II).

Based on the extremely low values of arterial O_2 content observed in normal, undisturbed, and submerged animals, Johansen (1966) suggested that lack of O_2 is not the primary stimulus inducing increased breathing efforts. In addition, he observed that the exposure of the animal to air was the only condition that "promptly instigates active air-breathing" in this species. However, Bicudo and Johansen (1979) and Heisler (1982) observed that when the water O_2 tension is reduced below 30 mm Hg, the animal switches from water- to air-breathing. In addition, Heisler (1982) showed that the arterial P_{O_2} levels in water-breathing and air-breathing *Synbranchus* are sufficient for more than 95 and 85% Hb saturation, respectively, despite the pronounced blood Bohr effect. Based on this tight regulation he suggested that P_{O_2} is likely the main factor in regulating the ventilation in this fish species.

The equilibrium and kinetic properties of the hemolysates of *Synbranchus* with O_2 and CO_2 are not exceptional (Phelps et al. 1979b). The magnitude of the Bohr effect for stripped Hb is about ($\phi = -0.135$), which is similar to that reported by Johansen (1970). In the presence of 1 mM ATP, however, this value increases up to ($\phi = -0.321$), indicating a strong modulatory effect of this organic phosphate on Hb–O_2 affinity. A large Bohr effect seems to be advantageous during air-breathing and it is therefore possible that this animal increases the intraerythrocytic levels of ATP during the air-breathing phase in order to increase the magnitude of the Bohr effect. No Root effect has been reported for this fish species.

4.2.2 The Stomach and Intestine as Air-Breathing Organs

Many loricarids and callichthids (see Table 4.4) from the Amazon use their vascularized stomach and intestine as an accessory air-breathing organ for gas exchange when water becomes severely deficient in O_2. As the amount of dissolved O_2 decreases, these animals start to surface rhythmically to swallow air. Gee (1976) extended this observation to members of the family Trichomicterydae. The mouth with its soft pursed lips is normally situated under, and a short distance back of, the snout in these fish species. Therefore, in order to breathe, the animal has to flip over and expose its ventral side to the water's surface, thus enabling it to take in a mixture of air and water by swallowing it through the mouth. Immediately after this, the animal dives back to the bottom of the water body. Once the O_2 has been removed, the air is released through the mouth, gill opening, or anus when the animal is surfacing or, more commonly, when the animal is in the water. Gradwell (1971a) described, in detail, this behavior in *Hypostomus regani* (= *Plecostomus punctatus*) using photographic analysis.

Pterygoplichthys multiradiatus, an armoured catfish from the family Loricariidae (known in the Amazon as bodó or acari-bodó), remains in the water near the bottom grazing for food during normoxic conditions and does not surface. During hypoxic periods, the animal surfaces rhythmically to gulp air, then returns to the original position. The entire cycle for this species lasts, on average, about 10 min although significant variations in air-breathing frequencies have been observed (Val et al. 1990). A similar behaviour was also observed in *Hypostomus regani*, a loricarid which is widely distributed across southern Brazil (Val et al. 1985). *Hoplosternum littorale* and *Corydoras aeneus*, armoured catfish of the family Callichthyidae, also exhibit similar behaviour (Gee 1976; Kramer and McClure 1981; Affonso 1990).

Most of these silurids are distinguishable from others by their extremely reduced swim bladder which is partially or completely encapsulated. It has been suggested that this reduction, resulting in negative buoyancy, has evolved either to promote a benthic life or to compensate for the increased buoyancy which results because of the gas held in the accessory air-breathing organ (see Gee 1976 for further details).

It is interesting to study in these fish species, namely *Pterygoplichthys* and *Hoplosternum*, their ability to survive long periods out of water. *P. mutiradiatus*, for example, can survive up to 30 h when exposed to air; after 12 h in air the animal is still in good shape, it becomes somewhat sluggish after 20 h (G.R. Bartlett, unpubl. data). About the same period out of water can be faced by *Hoplosternum* (Affonso 1990). In addition, it is very common in the Amazon to observe these fish being marketed alive, exposed to air for up to 30 h after their catch. No changes in blood O_2 content or Hb saturation have been observed during air exposure, indicating that the limiting factor for longer survival of these animals when exposed to air is not a reduced O_2 transfer.

Gee (1976) observed that the mean gas tension in the air-breathing organ is about 1.08 atm for all of the species that he analyzed under normal conditions. Gradwell (1971b) described a valve in the anterior part of the buccopharynx of some loricarids, including *Xenocara occidentalis*, *Hypostomus regani*, *Ancistrus annectens*, *Loricaria microlepidogaster*, and *Otocinclus mariae*, that is opened during inspiration and is closed during expiration preventing exhalation through the mouth. Thus, the used air in these fish species must leave through the anus. It is possible that this valve is present in many other loricarids. *P. multiradiatus* is a strong candidate to have this valve because this species displays a significant increase in the stomach/intestine gas tension when the anus is sutured. This increase was not observed when only the gill openings were sutured. Similar observations are also noted for *Hoplosternum littorale* (G.R. Bartlett, unpubl. data).

The air-breathing threshold varies widely among these fish species (Table 4.4). Interestingly, for some species, air-breathing occurs at all levels of dissolved O_2 while for the great majority, air-breathing occurs only when dramatic drops in the water O_2 levels are observed. However, all of them can survive long periods if denied access to the water's surface, even if the water is hypoxic. *Corydoras aeneus* and *Hoplosternum littorale* belong to the first group. These fish species breathe air regardless of the O_2 concentration of the water. They are capable of surviving long periods without access to the water's surface, even when the O_2 concentration is reduced to levels as low as 15 mm Hg for *Corydoras aeneus* and 20 mm Hg for *Hoplosternum littorale* (Kramer and McClure 1980; Affonso 1990). *P. multiradiatus*, *Hypoptopoma gulrae*, *Hemiodontichthys acipenserinus*, and *Dekeyseria amazonica*, among others belong to the second group (Val et al. 1990; Soares 1993).

Compared with the obligatory air-breathing fish, the Bohr effect of blood and stripped Hb in the loricarids and the callichthids are much less pronounced. For both *P. multiradiatus* and *H. littorale* a significant increase in the magnitude of the blood Bohr effect and a decrease in Hb–O_2 affinities were observed when they are exposed to normoxia relative to those exposed to hypoxia (Weber et al. 1979). This change can be attributed primarily to the changes in intraerythrocytic concentrations of organic phosphates which have a significant modulatory effect on the hemoglobins of these fish species (Weber and Wood 1979; Affonso 1990; Val et al. 1990). Under natural conditions, however, slight seasonal changes in both the magnitude of Bohr effect and the levels of organic phosphates have been observed for these fish species. It is important to note, however, that these changes show no correlation with the seasonal changes in dissolved O_2 concentrations in the water (Affonso 1990; Val et al. 1990). Interestingly, the Hb components of these two species can be divided, functionally, into two groups. In *P. multiradiatus*, one group contains the Hb fraction I which represents about 50% of the total Hb and is characterized by a discrete reverse Bohr effect; and the other group contains all other hemoglobins with normal Bohr effects (Brunori et al. 1979). In *H. littorale*, the Hb fraction I exhibits a reverse Bohr effect, which is reverted in the presence of 1 mM ATP, while the Hb fraction II exhibits a normal Bohr effect (Fig. 4.13). Both *P. multiradiatus* and *H. littorale* exhibit a moderate Root effect, while no Root effect has been observed for *Hypostomus* (Farmer et al. 1979).

The air-breathing ability in these species can be seen as an adaptation to low environmental O_2. However, at least in *P. multiradiatus*, the blood adjustments towards enhancing O_2 transfer to the tissues, commonly observed during hypoxia, are not offset by the air-breathing behaviour. In this species, air-breathing occurs simultaneously with lower intraerythrocytic GTP concentrations, higher hematocrit and Hb concentrations, and higher Hb–O_2 affinities (Val et al. 1990). Thus, at least in acari-bodó, the simultaneous occurrence of air-breathing enhancement of O_2 transfer from the blood indicates that the gas exchange through the stomach and intestine alone is not sufficient to meet the O_2 requirements of the animal.

With regards to CO_2 excretion, *Hypostomus*, at least, behaves as other facultative air-breathers, i.e., an increase in CO_2 retention occurs during hypoxia regardless of the opportunity for aquatic gas exchange (Wood et al. 1979b). Wood et al. (1979b) suggest that the gas exchange through the gills is reduced or even absent between air-breathing episodes and because the

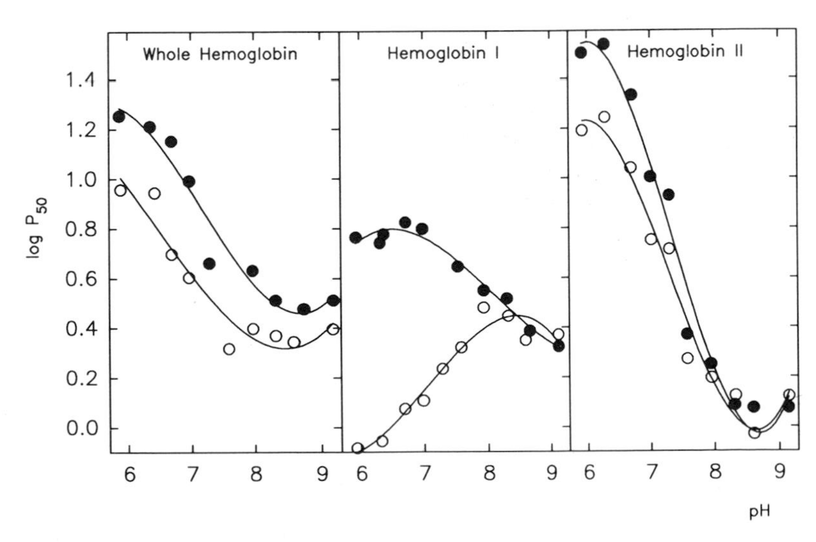

Fig. 4.13. The effect of pH and 1 mM ATP (*closed symbols*) on the O_2 affinity of unfractioned hemolysate, Hb fractions I and II of *Hoplosternum littorale*. (Redrawn from Garlick et al. 1979)

armoured skin does not allow CO_2 to diffuse out of the body, there is an increase in the partial pressure of this gas in the blood.

4.2.3 Swim Bladder as an Accessory Air-Breathing Organ

Air-breathing capacity has evolved independently in many unrelated lineages of teleosts (Liem 1987). Some of these lineages have secondarily developed the swim bladder as an air-breathing organ. This is seen in *Arapaima gigas* (Arapaimidae) and *Hoplerythrinus unitaeniatus* (Erythrinidae). However, not only is the structural organization of the swim bladder different between the two species but they exhibit a different behaviour, too. While *Arapaima* is an obligatory air-breather, *Hoplerythrinus*, known in the Amazon as jejú, is a facultative air-breather who uses its vascularized swim bladder for gas exchange only when there is a considerable drop in the concentration of dissolved O_2. In addition, *Hoplerythrinus* makes some overland excursions which have not been described for *Arapaima*.

Besides *Hoplerythrinus*, the Erythrinidae includes two other genera: *Hoplias*, with at least two water-breathing species, *H. malabaricus* and *H. lacerdae*, who are widely distributed across South America; and *Erythrinus*, with one species in the Amazon, *E. erythrinus*, which is a facultative air-breather. The main difference between them is the degree to which the swim bladder is vascularized (Fig. 4.14) and consequently the ability of the animal to extract O_2 from the air in the swim bladder. This ability is present in *Hoplerythrinus* but not in *Hoplias*, which cannot survive low levels of dissolved O_2 in the water. The gas bladder of *Hoplerythrinus* has an anterior portion which is connected to a posterior portion through a duct. The posterior portion is divided into respiratory (vascularized) and nonrespiratory (nonvascularized) sections (Kramer 1978).

When resting, *Hoplerythrinus* surfaces every 3 min for a breath of air when exposed to hypoxia. In contrast to the pirarucu, the jejú swims to the water's surface, moves its mouth out of water, and then opens it, taking in air which is pumped into the swim bladder. The ventilation per se is repeated two or three times while the animal is at the surface. This multiple ventilation per breath increases the O_2 tension in the swim bladder. This is important because, according to Kramer (1978), inspiration clearly precedes expiration in this fish species. Bubbles are not vented out when the animal is surfacing. Some bubbles leaving through the opercula, however, can be observed when the animal dives underwater. The gills are rhythmically ventilated when the animal is underwater and exposed to normoxia. However, gill ventilation becomes insignificant as the concentration of dissolved O_2 decreases (Farrell and Randall 1978; Kramer 1978; Stevens and Holeton 1978b). It is interesting to note that the movements of the mouth are very similar when breathing air and when breathing water. Kramer (1978) estimates that the mouth is opened for 84 ms when inspiring air while

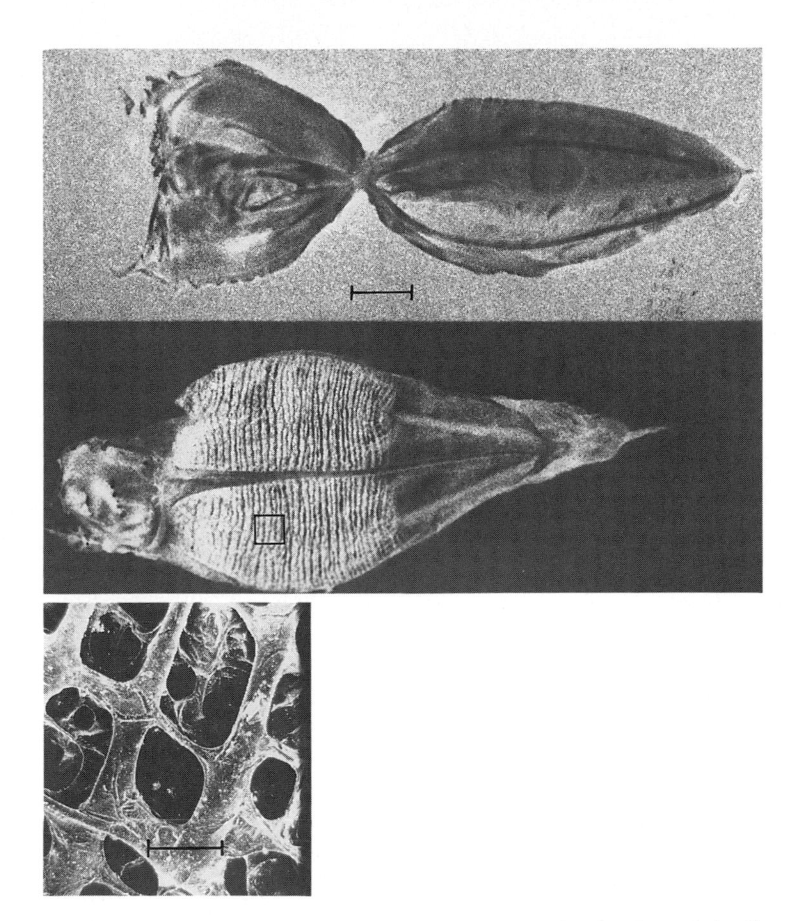

Fig. 4.14. The swim bladder of *Hoplias* (nonvascularized) and *Hoplerythrynus* (vascularized), *bar* = 10 mm. The *inset* shows a magnification (*bar* = 40 μm) of the vascularized structure of *Hoplerythrynus*. (Val, unpubl.; Randall, unpubl.)

it is open for 190 ms during water inspiration. However, mouth contraction lasts about 200 ms to pump air into the swim bladder and only 80 ms to pump water through the gills.

As the amount of dissolved O_2 decreases, the animal becomes more and more dependent on air-breathing to maintain its metabolic rate (see Fig. 5.5). Almost 100% of the O_2 is taken up from the air when the animal is exposed to water containing less than 20 mm Hg of dissolved O_2. Similar observations are valid for *Erythrinus* (Stevens and Holeton 1978b). The partial pressure of O_2 in the swim bladder rapidly decreases after the intake of air, this is an indication that O_2 is being transferred to the blood. In contrast, there are no changes in the partial pressure of CO_2, indicating that there is a reduction in the excretion of CO_2 through the swim bladder. Randall et al. (1978b) showed that nearly all CO_2 excretion occurs through the gills when the animal is in the water and that only 7% of the total CO_2 is excreted through the air-breathing organ. As a consequence, a significant increase in blood CO_2, simultaneous with a fall in pH, was also observed when the animal was exposed to air for 20 min. However, *Hoplerythrinus* can survive out of water for up to 24 h and so, there must be some CO_2 excretion through the skin or even through an increased participation of the swim bladder in this process when the blood P_{CO_2} tension is high. According to Stevens and Holeton (1978b), when the animal is put back into water after 24 h of air exposure, gill ventilation rate is not elevated, however, the depth of breathing is increased for about 1 h.

In *Hoplerythrinus*, a reduced blood perfusion of the gills arches 1 and 2 (these arches have well-developed secondary lamellae) takes place during air-breathing. Simultaneously, most or all of the blood is directed to the arches 3 and 4 (these arches have no secondary lamellae and very reduced filaments). Thus, in this way the fish is able to reduce the amount of O_2 leaked to the poorly oxygenated water during air-breathing (Farrell 1978; Smith and Gannon 1978; Stevens and Holeton 1978b). Also, as a consequence, a significant increase in blood CO_2 is observed during air-breathing because the great majority of this gas is removed via gills, at least when the blood tensions are normal (Randall et al. 1978b).

Lastly, one controversial point is the reduced renal acid excretion and percent of total ammonia excretion by the kidney in *Hoplerythrinus* when resting under normoxic conditions, as compared to *H. malabaricus* (Cameron and Wood 1978). These authors, to explain their findings, emphasized that the gills of *Hoplerythrinus* are completely functional and provide enough O_2 for the metabolic needs of the animal (as also in *Hoplias*) during water-breathing. However, they reported that the total lamellar surface area in *Hoplias* is 591 mm^2/g body weight, while in *Hoplerythrinus* it is only 251. This implies that the gills of the water-breather *Hoplias* are not as functional as in *Hoplerythrinus* with respect to the regulation of these parameters. Nevertheless, no significant differences in the activities of the enzymes of glycolysis, gluconeogenesis, pentose shunt, and hydrogen shuttling have been observed in the gills of these two species (Hulbert et al. 1978b). Thus, further studies are necessary to elucidate these aspects regarding these closely related fish species.

4.3 Water-Breathing

Although there are many obligatory and facultative air-breathing fish species, the great majority of the Amazonian fishes are water-breathers. In general, water- and air-breathers have faced the same environmental constraints. From an evolutionary point of view, it is interesting to observe that the air-breathing fish have developed many strategies to "buffer" the side effects induced by the air-breathing behaviour per se (predation, high blood

CO_2 levels, high O_2 availability, and O_2 leakage through the gills and skin, among others), while the water-breathers have developed many strategies to increase the O_2 transfer from the environment to the tissues and/or to avoid the problems caused by low environmental O_2 availability. These strategies, once again, include behavioural, morphological, anatomical, physiological, and biochemical modifications.

4.3.1 Avoiding Low Oxygen Conditions

Many water-breathing fish species are able to detect a current O_2 depletion in their environment or even to anticipate it, leaving the place before large drops in the water O_2 contents occur. According to Wootton (1990), this behavioural response to environmental changes is the fastest adaptive response. Many biochemical, physiological, and ecological mechanisms seem to provide the continuous information about the environmental conditions which are necessary, in order to enable the animal to choose between leaving or staying and facing the environmental constraint. It is important to point out that many species choose to stay despite an increased energy expenditure in searching for or adapting to low O_2 conditions. The main reason for remaining is because there is a reduction in the amount of competition for food as well as a decreased number of predators. In other words, the behavioural response seems to involve an evaluation of the environmental conditions as a whole, i.e., they weigh the advantages and disadvantages of leaving or staying and facing the unfavourable O_2 conditions.

Those fish species who do not have the ability to stay, migrate to other water bodies in search of better O_2 conditions. These migrations, as previously mentioned, involve small distances, normally between the river, where the water movements help to maintain the water's O_2 levels, and the floodplain areas, where significant diurnal and seasonal changes in water dissolved O_2 occur. This behavioral response is related to the resistance of the fish species to O_2 conditions. Junk et al. (1983), for example, studying the fish community of Lake Camaleão, a *várzea* lake of the Marchantaria Island, showed significant seasonal changes in species richness and abundance which were highly correlated with the seasonal variations in the water's O_2 content. These authors showed that the number of fish species at the station A increases significantly as a consequence of the impoverishment of the water conditions in the lake (stations B and C, Fig. 4.15). The types of species that tend to remain in the hypoxic lake are those who are specialized in some way to tolerate hypoxia, i.e., those species who are able to breathe air or who are resistant to hypoxia. It is possible that the diurnal lateral migrations described for many fish species of the Amazon are also influenced by the O_2 conditions in the lakes (Cox-Fernandes 1989).

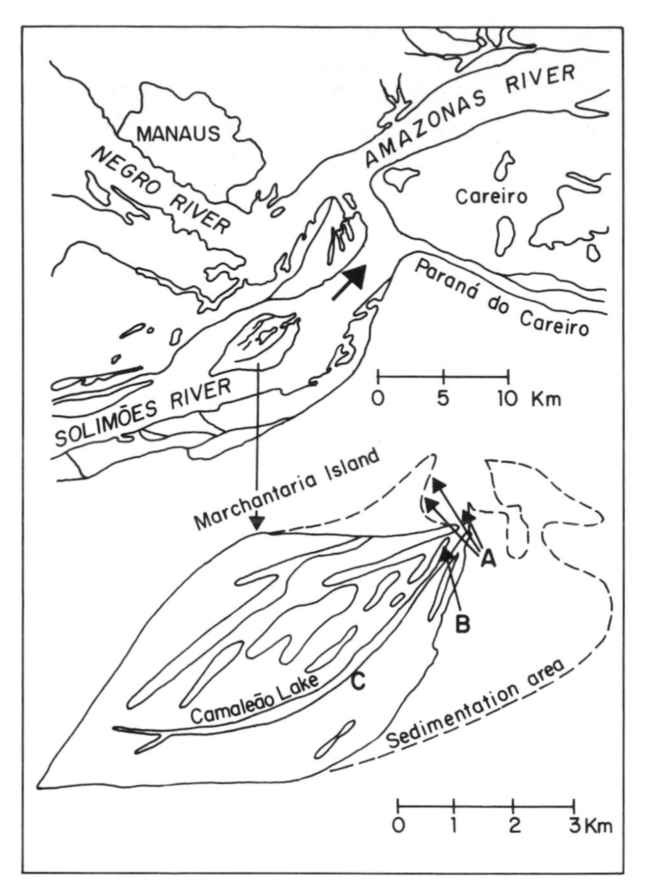

Fig. 4.15. Map of the Marchantaria island highlighting the main lakes and the collecting stations (*A*, *B*, *C*). (After Junk et al. 1983)

4.3.2 Skimming the Water's Surface

During hypoxia several water-breathing fish species of the Amazon skim the O_2-rich surface layers of water. This habit is known in the Amazon as *aiú*. Although appearing with different intensities and at different thresholds, skimming is observed in many nonrelated fish species, indicating the convergent character of this behaviour. These animals do not have an accessory air-breathing organ; they pump the O_2-rich layer of the water across the surface of the gills where gas exchange takes place as usual. Many fish species do not exhibit any particular morphological specialization in order to skim; however, several of them do exhibit a modification of the lower lip.

The capacity to modify the lower lip during hypoxia is best observed in *Colossoma macropomum* (Serrasalminae, Fig. 4.16) and *Brycon* cf.

Fig. 4.16. The Amazon tambaqui (*Colossoma macropomum*) with developed lips and the dorsal view of its head showing the inferior lip extended. (Val 1986)

erythropterum (Bryconinae) (Braum and Junk 1982; Braum 1983). This modification also appears in *Mylossoma duriventris*, *Mylossoma aureus*, *Piaractus brachypomum*, *Brycon* cf. *cephalus*, species of Triportheinae, among others. In both *C. macropomum* and *B. erythropterum* the expanded lip does not contain any blood vascularization, a clear indication that this structure does not have any gas exchange function. There is also no evidence for a gas storage function. The extension of the lip in these species seems to be the result of an increased lymphatic drainage to that region of the body when the animal is exposed to low O_2 in the environment (Braum 1983). Thus, the function of the developed lip seems to be purely mechanical, serving to improve skimming and consequently the uptake of O_2 by the gills.

When exposed to hypoxia in its natural environment, *C. macropomum* develops the inferior lip in about 2h. The frequency of individuals with a developed lip reaches 100% when the O_2 content of the water is below $2\,mg\,l^{-1}$ (Fig. 4.17). Interestingly, during the development of the lip, the animal also improves many biochemical and physiological modifications which buffer the negative effects of low O_2 concentrations in the water (Val 1986; Almeida-Val 1986; Almeida-Val and Val 1991a; Almeida-Val et al. 1993). These modifications include lower levels of erythrocytic ATP and GTP, increased anaerobic/aerobic metabolic rate, and increased hematocrit. After the lip is fully developed, there is a stabilization of plasma lactate levels and erythrocytic ATP and GTP (Val 1993; Almeida-Val et al. 1993). As the environmental O_2 is restored the lip disappears in about 2h.

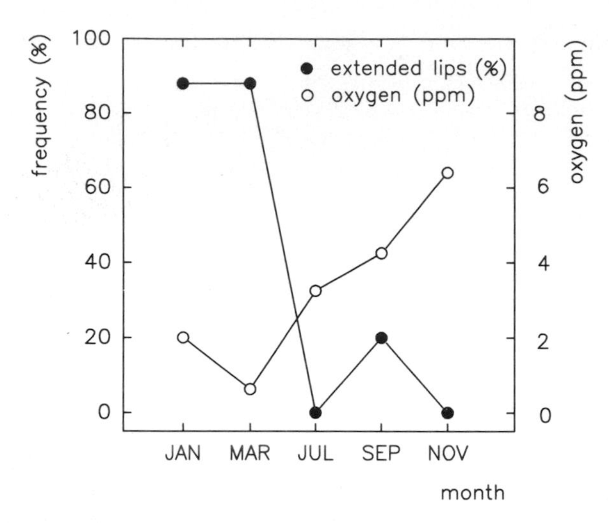

Fig. 4.17. Occurrence of *Colossoma macropomum* with developed lips and O_2 levels during different periods of the year 1984. (Val 1986)

According to Braum (1983), the modification of the lips in these fish species can be experimentally induced by reducing the O_2 concentration in the water to below $0.5\,mg\,l^{-1}$. By exposing *C. macropomum* and *B.* cf. *erythropterum* to this O_2 condition, the author described two distinct patterns of formation and disappearance of the lip. In *C. macropomum*, the modification of the lip occurs during the skimming and quickly disappears after the O_2 level is restored, while in *B.* cf. *erythropterum* the lip is already developed when the fish starts to skim and it disappears very slowly when the O_2 is restored. We believe that the physiological and biochemical determinants for the modifications of the lip are probably very similar in nature because the interspecific characteristics (i.e., modification of the lip and skimming behaviour) related to the threshold that induces each adaptive response are similar.

4.3.3 Improving Oxygen Transfer

Several fish species do not breathe air, do not skim the water surface, and do not leave their habitats when the level of dissolved O_2 drops. These fish species simply improve their ability to extract O_2 from poorly oxygenated environments by adjusting many biochemical and physiological mechanisms. These mechanisms are not mutually exclusive and do not occur with the same intensities in all species (Val 1993). Most of these mechanisms are very similar to those in fish species from the Temperate Zone. These mechanisms include regulation of the proportions of different Hb fractions, adjustment of intraerythrocytic levels of organic phosphates, changes in hematocrit, Hb concentration, and metabolic depression (Nikinmaa 1990; Randall 1990; Val et al. 1992a; Almeida-Val and Val 1993).

It is important to emphasize that these fish species, despite their ability to adapt to ambient O_2 concentrations which reach levels as low as $0.5\,mg\,l^{-1}$, cannot survive anoxic conditions and/or rapid changes in O_2 availability. Because these fish species represent a significant segment of the Amazon ichthyofauna, a sudden and/or drastic change in the environment, which causes a great reduction in the amount of dissolved O_2, induces a high mortality rate in fish. The best example is related to the phenomena known as *friagem*. During this period the temperature drops, inducing a complete circulation of the body of water, this causes the anoxic waters from the bottom of the water column to mix with and replace the O_2-rich surface waters. This causes the entire water column to become anoxic. As a consequence, there is a massive mortality of fish. Although factors other than anoxia, such as increased levels of hydrogen sulphide (H_2S), contribute to this mortality, lack of O_2 is the main cause. A massive mortality of fish was also observed when the Balbina dam started to operate. In this hydroelectric power plant, the water feeding the turbines and released downstream comes from the bottom of the lake which is almost anoxic.

In addition, many of the biochemical and physiological mechanisms developed in order to enhance O_2 transfer do not react rapidly enough to meet the oxygen demand of the animal when the environmental O_2 levels drop at a high rate. In other words, these transformations, in many species, are relatively slow. Hochachka and Somero (1984) and Wootton (1990) have shown that the lag times for biochemical and physiological reactions are considerably greater than that necessary for an adaptive behavioural response. Thus, in the case of the water-breathing fish of the Amazon not only the anoxia but the increased rate of O_2 depletion during those specific situations seems to have also contributed to the observed high mortality rates.

Many of the adaptive responses exhibited by the water-breathing fish of the Amazon when exposed to low O_2 conditions occur simultaneously, probably stimulated by the same factor. It has been shown that the catecholamines released into the circulation of fish in response to stressful conditions (hypoxia for example) play an important role in almost all processes related to O_2 transfer in fishes (Randall 1990). In fact, increases in the amount of circulating catecholamines have been related to the release of red blood cells from the spleen (Nilsson and Grove 1975; Perry and Kinkead 1989); control of red blood cell volume (Nikinmaa 1990); blood redistribution (Wood 1976); decreased erythrocytic levels of ATP (Milligan and Wood 1987); elevation of red blood cell pH (Primmett et al. 1986); and increased gill epithelial permeability (Isaia et al. 1978), among others. Of course, these effects cannot be generalized because different fish species have accumulated different genetic characteristics throughout the evolutionary process.

4.4 The Blood

Blood is a very specialized tissue located at the interface between the organism and the environment. The study of the relationship between blood and environmental characteristics began with the work of Krogh and Leitch (1919), receiving increased attention ever since. The blood of organisms facing drastic environmental conditions such as the fish of the Amazon, however, has been hardly studied.

Fish blood contains three different groups of cells: the nucleated erythrocytes which are involved with the transport of O_2 and CO_2; the thrombocytes and/or platelets which are involved with hemostasis; and the white cells (lymphocytes, macrophages, monocytes, basophils, eosinophils, and neutrophils) which are components of the immune system. Thrombocytes, platelets, and white cells in fishes of the Amazon have been addressed and understood in only a few studies.

A significant inter- and intraspecific variation of circulating white cells have been described in fishes of the Amazon. Yano (1989) and Moura (1990) observed that the intraspecific variation was related to the catch site; i.e., the animals collected in the *igarapés* crossing Manaus contained higher levels of white cells than the animals collected in the floodplain areas and main rivers.

4.4.1 The Erythrocytes

The nucleated erythrocytes are the dominant type of cell in the blood of the most primitive to the most specialized fishes of the Amazon. This pattern has been observed in the vast majority of fishes, except, of course, in the icefish. The number and volume of these erythrocytes, however, exhibit wide intra- and interspecific variations among fishes. In general, there is an increase in the number of red blood cells as the degree of specialization of the vertebrates increases, however, this relationship is weakened when a small number of species within the same group is analyzed (Fig. 4.18). In fact, Myxine and Chondrichthyes have a smaller number of erythrocytes than the Osteichthyes. However, there is no clear relationship between the number of erythrocytes and the phylogenetic position of the different orders of Osteichthyes.

On the other hand, the intraspecific variation in the number of red blood cells observed for fishes of the Amazon is strongly related to the environmental conditions of the habitat and to the biological conditions of the animals (Amadio 1986; Monteiro et al. 1987; Affonso 1990; Salvo-Souza 1990). For example, the number of circulating erythrocytes is related to sex, sexual development, size, and migration phase in species of the genus *Semaprochilodus* (Val 1983; Amadio 1986). In *Arapaima gigas*, Salvo-Souza (1990) observed a significant decrease in the amount of circulating

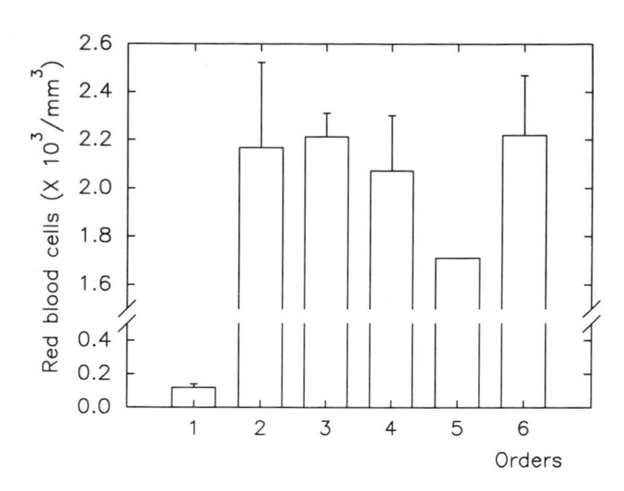

Fig. 4.18. Number of circulating red blood cells of selected orders of fish. The orders are organized along the abscissa from the generalized to the specialized according Nelson (1984). *1* Rajiformes – not Amazonian; *2* Osteoglossiformes (n = 3); *3* Characiformes (n = 23); *4* Siluriformes (n = 14); *5* Synbranchiformes, not Amazonian (n = 1); and *6* Perciformes (n = 8)

erythrocytes as the animal grows. However, stronger relationships have been observed between the number of erythrocytes and environmental parameters, namely, the amount of dissolved O_2 and variations in temperature, for many fish species of the Amazon (Val 1993). Indeed, these relationships are strongly evident when those environmental parameters are artificially controlled in the laboratory. Under natural conditions, a strong correlation with water level oscillation is observed. The fluctuations in the water level of Amazonian waters is an environmental factor that encompasses the interrelationships between many biotic and abiotic characteristics of the environment. Thus, it is not surprising that many physiological parameters exhibit a strong correlation with water level oscillations but not with an individual environmental parameter such as temperature or dissolved O_2.

Monteiro et al. (1987), for example, showed a significant increase in the number of circulating red blood cells of *Mylossoma duriventris* during periods when the water level was low, despite the higher concentration of dissolved O_2 and higher temperatures. Their explanation of this finding involves the coexistence of a high number of fish species and specimens who increase the competition for food. Consequently, the predation rates must increase, directly affecting the metabolic activity of the animals. In addition, the peak levels of dissolved O_2 measured around noon, do not reflect the O_2 dynamics of the water tract, i.e., the diel oscillation during the times when water levels are low, is much more intense than during periods when water levels are high. This means that during seasons when the water is low, it is common to observe very low levels of O_2 or even anoxic conditions during night. Therefore, the higher number of circulating red blood cells during this season may reflect an adaptive response to the diurnal O_2 regime. No significant changes in the number of circulating red blood cells were observed, however, for *Prochilodus* cf. *nigricans* when facing similar environmental conditions (Val et al. 1992a). On the other hand, despite the significant differences that she observed among the mean number of red blood cells in *Hoplosternum littorale*, who were studied under natural conditions, Affonso (1990) failed to show a correlation between the number of red blood cells and temperature, the concentration of dissolved O_2, or the water level.

The increase in the number of circulating erythrocytes has been correlated with the effect of released catecholamines on the spleen (Nilsson and Grove 1975; Perry and Kinkead 1989). As a consequence of spleen contraction, immature erythrocytes are also released into the circulation. Moura (1990) reported an increase in the number of these immature cells, including proerythrocytes, in specimens of *Hoplosternum littorale*, after they were exposed to stressful conditions. These immature red blood cells are, in general, bigger than the mature erythrocytes (Fig. 4.19) and may explain, in part, the modification of the relationship between number of red blood cells and hematocrit after the animal has been exposed to a stressful condition. It is interesting to observe that these immature red cells normally contain a number of organelles (Golgi complex, mitochondria, ribosomes, centrioles, among other structures) which are lost in the mature form. As a consequence, a progressive increase in the amount of Hb is observed during the red cell maturation process (Rowley et al. 1988). Thus, the new released red blood cells have a smaller O_2 carrying capacity than the mature ones.

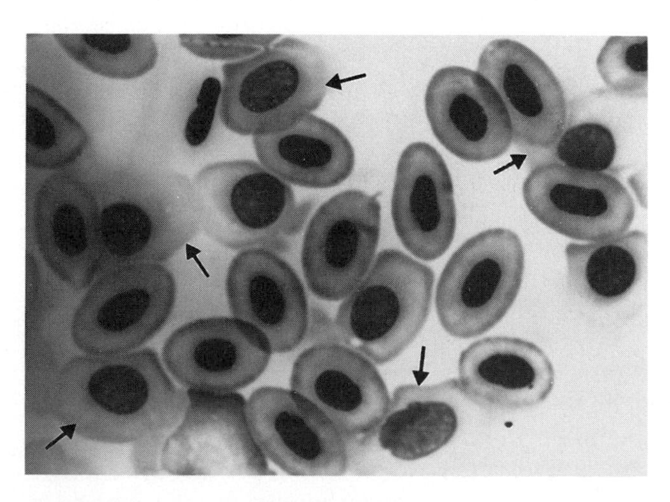

Fig. 4.19. Immature erythrocytes of *Hoplosternum littorale* (*arrows*). Note the smaller size of the mature cell. (Moura and Val, unpubl.)

Hematocrit changes can also be a result of changes in the volume of the red blood cells. Many environmental and physiological factors have a direct effect on the delicate acid–base balance of the blood which directly affects the red blood cell volume (see Nikinmaa 1990 for further details). This seems to be the case for many fish species of the Amazon which exhibit a significant change in hematocrit values but no changes in the number of circulating red blood cells. Consequently, a significant change in mean corpuscular volume (MCV) is observed (ALV unpubl. data). In addition, there is a general trend towards increased MCV values in species with sluggish habits and adapted for aerial respiration.

In general, higher values for hematocrit, Hb concentration, and the number of circulating red blood cells have been seen in species exhibiting rapid movements and predatory behaviours compared to those exhibiting sluggish and nonpredatory behaviours. Similar observations have been reported for fish species from other environments and other climate zones (Coburn and Fischer 1973; Belokopytin and Rakitskaya 1981; Gutiérrez and Sarrasquete 1983; Rambhaskar and Rao 1987). However, when closely related water- and air-breathing fish species are compared (*Arapaima* and *Osteoglossum*, *Hoplerythrinus* and *Hoplias*, for example) a reverse trend is observed (Johansen et al. 1978a).

4.4.2 The Hemoglobins

The hemoglobin is composed of two main parts: the globin and the haem unit. The haem unit is identical in all fish species studied thus far, but the globin, which is a protein, differs from species to species (see Chap. 3). The different Hb fractions present in one fish species were termed isohemoglobins by Powers (1974), an analogy to terminology used for enzymes. In general, each Hb molecule includes four globin chains, normally two α and two β, and four haem units which bind reversibly with O_2 and CO_2. The great majority of fish species have symmetric hemoglobins, i.e., two pairs of identical globin chains present in each Hb. Some fish species, however, have asymmetric hemoglobins, i.e., at least three different globin chains are present in one Hb molecule. This is the case for *Pimelodus maculatus* (Reischl and Tondo 1974). Another interesting aspect of the structure of teleost hemoglobins is the presence of only two $-NH_2$ termini per tetramer, compared with four in mammals. These $-NH_2$ are located at the β chains. As a direct consequence, there is a reduced sensitivity of fish hemoglobins to CO_2 (Riggs 1979).

4.4.2.1 Blood Levels and Solubility of Hemoglobins

The concentration of Hb in the blood of fish varies greatly. Values ranging from 4–5 $g\,dl^{-1}$ up to 20–21 $g\,dl^{-1}$ are very common. This variation occurs both inter- and intraspecifically. In fishes of the Amazon, the intraspecific variation is primarily related to changes in water levels and dissolved O_2 levels, and secondarily to sex, sexual development, size, feeding behaviour, respiration habit, and migration phase, among others (Val 1983; Amadio 1986; Affonso 1990; Salvo-Souza 1990; Val 1993). In general, there is a significant relationship between the blood Hb concentrations and hematocrit (Val 1986; Affonso 1990; Salvo-Souza 1990). Interestingly, no significant differences are observed for the regression lines between these parameters for obligatory, facultative, and water-breathing fish species (Fig. 4.20). When the average of the Hb concentration and the hematocrit of different fish species are plotted together, there is a general trend towards lower values for both parameters in water-breathers and higher values in air-breathers (Fig. 4.20).

The importance of the Hb in transporting O_2 from the environment to the tissues has been justified by many authors in terms of cardiac output. Satchell (1991), for example, estimated that if only plasma was used to transfer O_2 to the tissues in a 1-kg trout, the cardiac output would have to be four times greater than current values. According to him, this is an impossibly high value. However, many species, including rainbow trout and many Amazonian species, survive very low Hb concentrations and hematocrit values (Steen and Berg 1966; Iwama et al. 1987; Affonso 1990; Val et al. 1994; ALV, unpubl. data). In addition, Iwama et al. (1987) showed that plasma adrenaline and noradrenaline increased significantly only when the hematocrit was experimentally reduced to below 7.8%. Thus, it is possible that even major changes in hematocrit and Hb concentration, between certain limits, will not represent a major source of stress, at least

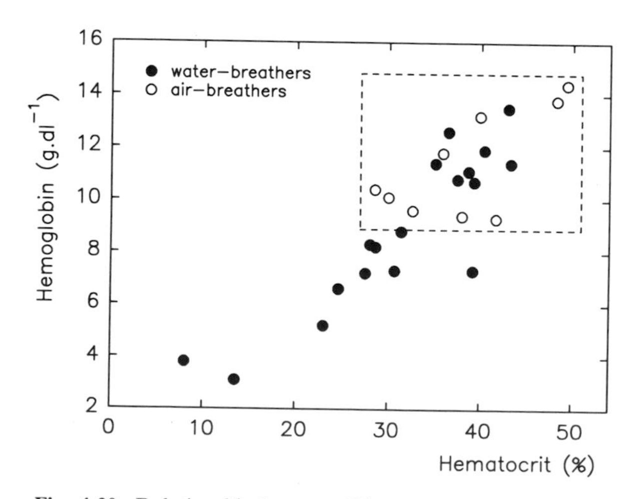

Fig. 4.20. Relationship between Hb concentration and hematocrit in water- and air-breathing fish of the Amazon. Note the higher mean values of Hb and hematocrit in air-breathers. Each point represents one species. (Data from Salvo-Souza and Val, unpubl.)

for animals under resting conditions and in darkened Perspex boxes. However, it was found that the blood's carrying capacity for O_2 is significantly affected when levels are below or above the optimum hematocrit value (Wells and Weber 1991) and also that anaemic fish apparently adjust some physiological parameters in order to maintain O_2 delivery to the tissues (Val et al. 1994).

So, why do many fish species maintain very high levels of Hb? Numerous reasons can be evoked to answer this question. However, two reasons appear to play a major role. The first is the sporadic or even constant exposure to low O_2 levels associated with an oscillating metabolic demand for O_2 in animals exposed to their natural environmental constraints, which include regular exposure to predation pressures and searching for food. Secondly, it is possible that under the natural oscillating environmental conditions there is an increased participation of the hemoglobins in buffering the blood acid–base disturbances. In addition, the evolutionary pressures seem to have been directed towards preserving, rather than reducing blood Hb concentration.

According to Riggs (1976), Hb concentration is at the limit of its solubility in the red cell. Moreover, the deoxyhemoglobin is less soluble than its counterpart, the oxygenated form (Riggs 1979). For fish living in an environment with unstable O_2 levels, any mutation decreasing the solubility of the Hb would be a disadvantage. This seems to be the reason for an increase in the number of Hb fractions occurring simultaneously with the preservation of Hb concentration, observed in fishes of the Amazon. There is, however, no clear relationship between the number of Hb fractions and environmental stability (Fyhn et al. 1979; Perez and Rylander 1985; Val et al. 1987; Val 1993).

4.4.2.2 Adjustments of Hemoglobin Proportions

Despite the characteristics mentioned in Section 4.4.2.1, the animals having multiple Hb systems are able to adjust the relative concentration of each Hb fraction according to the season, age, and amount of dissolved O_2 (Val 1986; Monteiro et al. 1987; Val et al. 1990). For example, the relative concentration of the four Hb fractions in the adult *Colossoma macropomum* are different during lowest water level periods when compared to highest water level periods (Val 1986). The hemoglobins of *Mylossoma duriventris* exhibit similar behaviours (Fig. 4.21). In *Pterygoplichthys multiradiatus*, who are acclimated to normoxia and hypoxia, a slight but significant difference was observed for the Hb fractions I and III, although no significant differences were observed for Hb fractions II and IV.

Semaprochilodus insignis, on the other hand, possesses an intriguing Hb system. This fish species has four Hb fractions (Fig. 4.22). The Hb fraction II is the largest component and does not exhibit any significant intrapopulational variation or variations due to the migration phase of the animals.

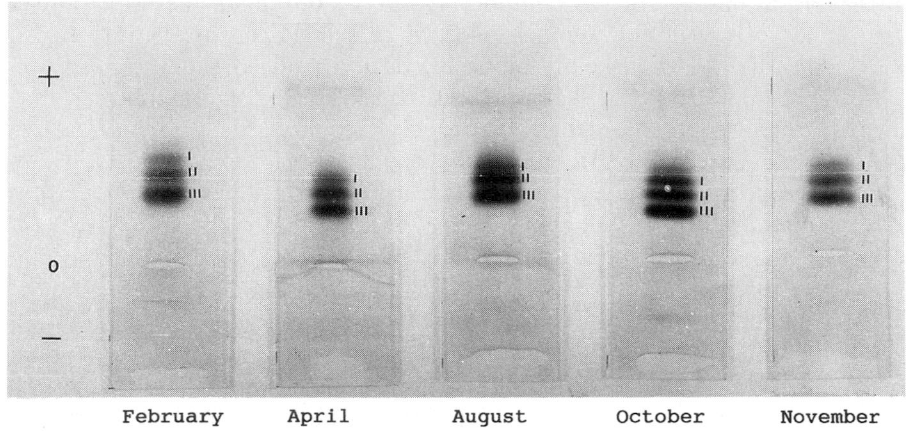

	February	April	August	October	November
I	19.64±1.55	23.62±2.53	22.73±2.99	21.73±1.87	17.39±2.69
II	33.38±1.03	36.63±1.63	33.71±1.83	33.03±1.19	36.17±1.52
III	46.98±2.15	39.76±3.05	43.53±2.94	45.23±2.11	46.50±2.75

Fig. 4.21. Electropherograms of hemoglobins of *Mylossoma duriventris*, obtained on agar-starch coated slides, during different periods of the year. The relative concentrations of the Hb fraction are indicated. (Monteiro et al. 1987)

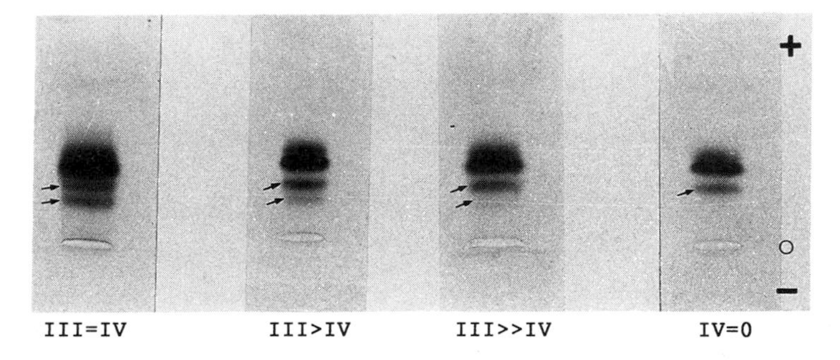

Fig. 4.22. Electropherograms of hemoglobins of *Semaprochilodus insignis* obtained on agar-starch-coated slides. The *arrows* indicate the differences in the concentration of the Hb components III and IV. These phenotypes occurs simultaneously in the population. (Val et al. 1986)

Fraction I is the second largest and exhibits a significant variation corresponding to the migration phase of the animals. Fractions III and IV show the greatest intrapopulational variation; fraction IV is absent in some individuals, in others it has the same relative concentration as fraction III, and in most individuals fraction III has a higher relative concentration than fraction IV (Val et al. 1986). This pattern of intrapopulational variation can be seen as a result of either polymorphic regulatory genes or as individual adjustments. In the first case, a polymorphism in the regulatory gene of one globin chain common to fraction III and IV could explain the interrelated changes in their relative concentration. In the second case, the synthesis of this globin would be related to individual characteristics. Further studies are necessary to elucidate this kind of regulation. However, this case clearly indicates that the adjustments of the proportion of different Hb components have a strong individual component either genetic and/or physiological. In addition, this kind of regulation may have contributed to the absence of a correlation between number of Hb fractions and environmental characteristics.

4.4.2.3 Why So Many Hemoglobin Fractions?

Besides the fact that an increase in the number of Hb fractions increases their solubility, it has been shown that different Hb fractions have different properties in many fish species (Brunori et al. 1979; Bunn and Riggs 1979; Garlick et al. 1979; Riggs 1979; Val et al. 1986). Bunn and Riggs (1979), analyzing the hemoglobins of 16 fish species from the Amazon by electrofocussing, showed that six have at least two Hb fractions, one of which becomes more acidic upon deoxygenation, while the others become less acidic. This means that, for these fish species, at least one Hb fraction has an increasing Hb–O_2 affinity as the pH rises, whereas the others have a lower O_2–affinity as the pH rises. We tried to relate this feature with the characteristics of the habitat of these fish species but we failed to show any correlation. Other authors have confirmed Bunn and Riggs' findings by analyzing many species individually using classical methods (Brunori et al. 1979; Garlick et al. 1979).

Based on these facts, many authors have referred to the adaptive character of multiple hemoglobins. However, once again, no clear correlation exists between the number of species having multiple hemoglobins with different functional properties and environmental characteristics such as dissolved O_2, water pH, temperature, animal behaviour, presence of choroid rete, or swim bladder (Riggs 1979; Val et al. 1987). Powers (1972) and Riggs (1976) suggested that the presence of many different hemoglobins can provide an extra protection against the lethal effects of hyperactivity but as far as we know, this relationship has yet to be shown to exist in natural populations.

In summary, many species have several different types of hemoglobins whose relative proportions may be regulated according to environmental and physiological characteristics. A significantly high proportion of these species have at least one Hb fraction with quite different functional properties. No correlation exists between these characteristics and environmental and/or physiological parameters. Thus, the question of why some fish have multiple hemoglobins remains to be answered.

4.4.2.4 Methemoglobin Levels

Methemoglobin, in contrast to Hb, is unable to bind O_2 reversibly because its iron (Fe) is in the ferric (Fe^{3+}) form rather than in the ferrous (Fe^{2+}) form. Increases in methemoglobin levels result from auto-oxidation of Hb combined with reduced activity of the enzyme methemoglobin reductase. No significant difference exists between the methemoglobin levels in fish from temperate and tropical zones, which varies between 2 and 21% of total Hb (Cameron 1971; Bartlett et al. 1987; Brauner et al. 1993). Increases in Hb oxidation rates can be observed, for example, in fish exposed to water containing high levels of nitrite. Nitrite is found in both natural aquatic systems and aquacultural facilities. In natural systems, the nitrate, which is washed from agricultural fertilizers, is anaerobically reduced to nitrite while its accumulation in aquacultural facilities with reduced water flow is due to the bacterial oxidation of ammonia to nitrite (Collins et al. 1975). Reduced activity of methemoglobin reductase in animals results mainly from an inherited defect in its structural and/or regulatory genes (see Schwartz et al. 1983 for further details). The effect of increased methemoglobin levels on fish is the same as in man: hypoxia.

Studying the effect of nitrite on fishes of the Amazon, we have reported an extensive, interspecific variation in the levels of methemoglobin after 1 h of nitrite administration (Bartlett et al. 1987). The observed interspecific variation is neither related to the respiration habits of the animals nor to the hematocrit or Hb concentration. The variation is probably related to fluctuations in the activity of the NADH-dependent enzyme, methemoglobin reductase and/or in the blood levels of some specific chemical compounds that convert methemoglobin back to the functional Hb, such as vitamin C. In this study we also found, using in vitro samples, that a low pH (below 6.5) has a significant effect on the rate of auto-oxidation of hemoglobin to methemoglobin in *Semaprochilodus insignis*. However, there is no apparent relationship between methemoglobin levels and water pH (ALV, unpubl. data).

4.4.3 Oxygen Affinity of Hemoglobins

When the partial pressure of O_2 in the blood is plotted against the amount of O_2 bound to the Hb, the result is generally a sigmoidal curve known as an oxygen dissociation curve. This curve is shifted to the left or right in response to the changes in many physiological parameters which includes pH, levels of allosteric modulators, as well as some anions and cations. The O_2 affinity of the Hb is conveniently represented by the P_{50} (i.e., the partial pressure of O_2 at which the Hb is 50% saturated with O_2). By plotting the $\log P_{50}$ (the higher the P_{50}, the smaller the $Hb-O_2$ affinity) versus pH, a good picture of the dynamics of the O_2 transfer is obtained. In general, for the whole hemolysate or for the whole blood, a normal Bohr effect is observed, i.e., the $Hb-O_2$ affinity increases as the pH increases. However, for some specific Hb fractions, a reverse situation is observed.

The stripped $Hb-O_2$ dissociation curve is an intrinsic characteristic of the molecule and depends primarily on the globin chain composition, whose interaction will dictate the ease with which the O_2 will bind to the second and third haem groups (cooperativity, indicated by Hill's coefficient). Thus, this is an inherited characteristic that cannot be changed during the life span of the individual. Indeed, if the hemolysate contains more than one type of Hb with different functional properties, the final shape of the O_2 dissociation curve will be a function of the concentration of each Hb fraction in the hemolysate. This is the case for many fish species because many of them have hemolysates containing more than one type of Hb whose concentrations are continuously adjusting.

The effect of pH on the O_2 affinity of hemolysates is known for many fish species of the Amazon, including *Lepidosiren paradoxa*, *Arapaima gigas*, the stingray *Potamotrygon*, *Mylossoma* spp., *Loricariichthys* spp., species of the genus *Semaprochilodus*, *Serrasalmus rhombeus*, and species of the genus *Brycon* (Martin et al. 1979a,b; Phelps et al. 1979a; Galdames-Portus et al. 1979; Wood et al. 1979c; Val et al. 1986; Val and Almeida-Val 1988). For many of them, this effect also occurs in isolated Hb fractions. The characteristic sigmoidal shape has always been observed with rare exceptions. For example, this characteristic shape is observed for *Brycon* cf. *erythropterum* but not for *B.* cf. *cephalus* (Fig. 4.23). Interestingly, the Hb electrophoretic

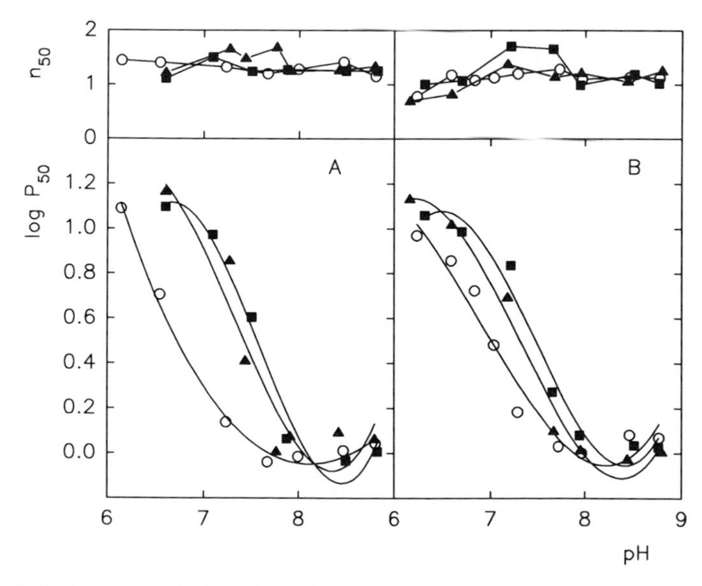

Fig. 4.23. The effect of pH (*open circle*), ATP (*closed triangle*), and GTP (*closed squares*) on the O_2 affinity and cooperativity of unfractioned hemolysate of **A** *Brycon* cf. *cephalus* and **B** *Brycon* cf. *erythropterum*. (Redrawn from Val and Almeida-Val 1988)

patterns for these species differ only by two secondary fractions and the presence of ATP or GTP restores the characteristic shape of the curve (Val and Almeida-Val 1988).

In general, the interactions between globin chains to facilitate O_2 binding is very low in fish hemoglobins. In the fishes of the Amazon, this also seems to be the case since cooperativity values above 2.5 have not been reported. In general, values close to one are observed without any significant effect of pH. However, Galdames-Portus et al. (1979), studying the hemoglobins of *Osteoglossum bicirrhosum* and *Arapaima gigas*, reported a significant effect of pH on Hill's coefficient. In *Osteoglossum* these authors observed a peak for cooperativity values at around pH 7 for stripped Hb and around pH 7.5 for Hb plus ATP. In *Arapaima*, increased cooperativity values are observed when the pH rises above 7.5 without a significant effect of ATP. Effects of pH on Hill's coefficient were also reported for *Lepidosiren paradoxa* (Phelps et al. 1979a). It is interesting to observe the reverse behaviour of the stripped hemoglobins of the two obligatory air-breathers (Fig. 4.24).

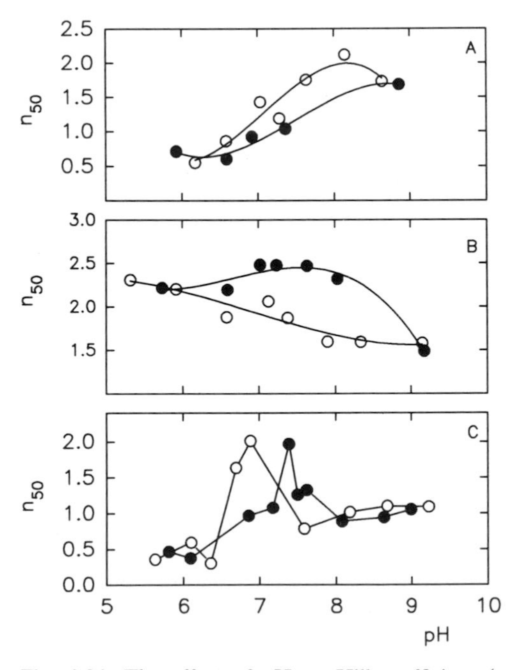

Fig. 4.24. The effect of pH on Hill coefficient (n_{50}) of **A** *Arapaima*; **B** *Lepidosiren*; and **C** *Osteoglossum*. *Closed symbols*, 1 mM ATP; *open symbols*, no ATP. (Data from Phelps et al. 1979a and Galdames-Portus et al. 1979)

4.4.3.1 *Effect of Temperature on Hb–O_2 Affinity*

As the water temperature increases, fishes face two problems: the decreased solubility of O_2 in the water and the increase in their biological and metabolic activities (see Chap. 5). These fish must, therefore, increase their O_2 uptake from the environment even though there is less O_2 available. Johansen and Lenfant (1972) hypothesized that species experiencing relatively constant thermal conditions will have hemoglobins which are quite sensitive to temperature changes. Conversely, the animals living in areas with large thermal fluctuations will display little sensitivity. This hypothesis tends to hold true for fish in the Amazon where a significant effect of temperature on Hb–O_2 affinities has been observed for a variety of species without any apparent relationship with phylogenetic status, respiratory habit, and/or conditions of the habitat (Powers et al. 1979; Val 1986). However, fish such as *Fundulus* who experience large thermal fluctuations contain hemoglobins which are equally sensitive to temperature (Fig. 4.25). Thus, Powers et al. (1979) suggest that the generalization proposed by Johansen and Lenfant (1972) is not justified.

It is interesting to observe that, despite the constant annual temperature averages, significant diel oscillation occurs. In addition, significantly decreased temperatures occur during the *friagem* period (see Chap. 2). The diel oscillation pattern is well defined with the highest temperatures occurring simultaneously with the highest solar incidence. Thus, the decreased Hb–O_2

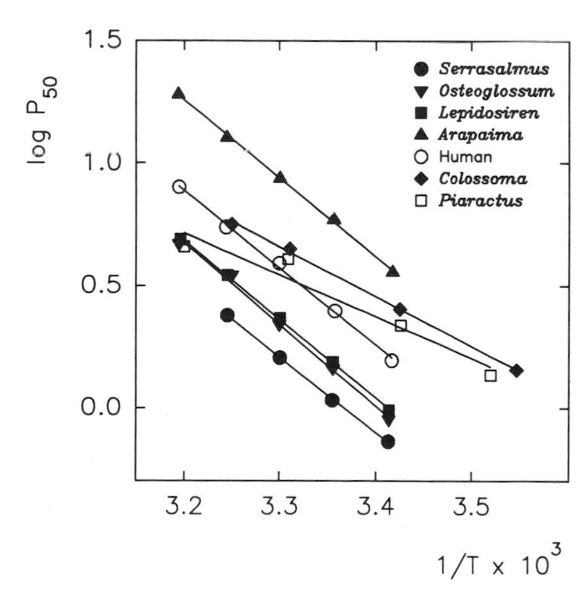

Fig. 4.25. Temperature effects on Hb–O_2 affinity in selected fish species of the Amazon. Human hemoglobin (Hb A) has been included for comparison. (Data compiled from Powers et al. 1979 and Val 1986)

affinity observed by Powers et al. (1979) and Val (1986) at high temperatures occurs when the fish have the highest amount of O_2 available to them. Thus, it is quite possible that the temperature regime in the Amazon does not represent an evolutionary pressure that is strong enough to change the temperature sensitivity of fish Hb and this is why they have kept the characteristics of their ancestors.

4.4.3.2 The Effect of Phosphates on Hb–O_2 Affinity

In general, the organic phosphates, particularly ATP and GTP, bind to fish deoxyhemoglobin, reducing its O_2 affinity. The binding sites of organic phosphates for fish hemoglobins has been determined for a few species (reviewed by Nikinmaa 1990), including the Amazonian lungfish (*Lepidosiren paradoxa*), who were studied by Rodewald et al. (1984). The binding site seems to be compatible for the binding of both ATP and GTP, however, GTP is a stronger allosteric effector than ATP. The effect of the organic phosphates on Hb–O_2 affinity is a function of two factors: the negative charge on the organic phosphate (i.e., the greater the charge, the greater the effect), and the structure of the binding site. Gillen and Riggs (1977) found that an addition of 1 mM ATP has the same effect on O_2 affinity of a variety of teleost fish hemoglobins, as a drop in pH of 0.5 units. For IHP, this ratio is 0.7 mM : 1.6 unit drop in pH. The effect of GTP lies between these extremes.

The effect of these allosteric modulators on O_2 affinity increases as the pH decreases. In general, similar P_{50} values to those for stripped Hb are observed at an alkaline pH. This value increases significantly as the pH becomes more acidic. If the structure of the binding site allows for only two hydrogen bonds, the effect of GTP is similar to that of ATP. Otherwise, the effect of GTP is more pronounced than that of ATP. This is the case in *Brycon* cf. *erythropterum* (Fig. 4.23). One can note that the addition of ATP and/or GTP does not induce a significant change in the cooperativity between globin chains in this fish species. In other species such as *Lepidosiren paradoxa*, *Arapaima gigas*, and *Osteoglossum bicirrhosum* a significant cooperativity change is observed (Galdames-Portus et al. 1979; Phelps et al. 1979a).

For some hemoglobins, such as the Hb I in *Hoplosternum littorale*, the effect of ATP is so strong that it restores the characteristic shape of the Bohr plot (Garlick et al. 1979; see Fig. 4.13). Interestingly, this strong effect only occurs when the pH is below 8.3. At pH 8.3 and above, this effect is reduced and the P_{50} values are similar to those of stripped Hb. As in *Brycon* no significant effect of ATP on cooperativity values is observed.

The effect of organic phosphates on O_2 affinity is not restricted only to their direct binding with Hb. As their total intraerytrocytic concentration increases, a significant drop of intracellular pH is observed, consequently increasing the Hb–O_2 affinity, at least for the hemoglobins exhibiting normal

Bohr behaviour. This nonspecific effect of organic phosphates seems to be important for some fish species such as *Anguilla anguilla* (Wood and Johansen 1973) and should be evaluated in species facing high temperatures such as the Amazonian fishes.

4.4.4 Oxygen Affinity of Blood

The blood O_2 affinity, on the other hand, is a function of five main parameters which are: (1) the intrinsic O_2 affinity of the Hb, (2) the concentration of Hb within the red blood cell, (3) the sensitivity of Hb–O_2 affinity to heterotropic ligands, (4) the concentration of these ligands within the red blood cells, and (5) temperature (see Nikinmaa 1990 for further details). Thus, by changing these parameters the animal can change the blood O_2 affinity, adjusting the O_2 transfer according to the physiological needs and/or to the environmental conditions. In other words, the animal can adaptively adjust the blood O_2 affinity.

Val et al. (1992a), studying *Prochilodus* cf. *nigricans*, showed that its blood O_2 affinities varied according to the intraerythrocytic levels of GTP and GTP/ATP ratios. The lowest blood O_2 affinities were observed in November when GTP and GTP/ATP ratio were high. The highest blood O_2 affinities were observed in July, simultaneously with the lowest levels of GTP and GTP/ATP ratios. Intermediate values for these parameters occurr in October. Val et al. (1990) observed a tendency for higher blood O_2 affinities, due to changes in intraerythrocytic levels of organic phosphates, in specimens of *Pterygoplichthys multiradiatus* (a facultative air-breather), which were exposed to hypoxic conditions. Affonso (1990) also observed a significant seasonal change in the blood O_2 affinities of another facultative air-breather, *Hoplosternum littorale*. These modifications reflect the environmental temperature changes and are in response to variations in the concentrations of erythrocytic phosphate.

When analyzing a small number of closely related species, a significantly higher blood O_2 affinity is observed for air-breathing fish species. Johansen and Lenfant (1972) were the first to suggest that the blood of air-breathing fish is adapted to release O_2 at higher partial pressures than is the blood of their fellow water-breathers. However, Powers et al. (1979), studying blood O_2 affinities of fishes belonging to 40 different genera from the Amazon, reported some controversial points. For example, these authors failed to show any correlation between blood O_2 affinities and respiratory behaviour. Instead, they showed that the blood O_2 dissociation curves of air-breathers are included in the range of those for water-breathers. They also showed that fish inhabiting "rapid waters" possess significantly higher P_{50} values when compared to those inhabiting "slow water". Almeida-Val et al. (1985) established that the O_2 dissociation curves for seven migratory fish species of anostomids lie within the boundaries set by the curves for fish which swim in rapid river zones and fish that swim in slow moving water zones (Powers et al. 1979). This sort of comparison seems to be much more reliable for close related species because blood O_2 affinity is a function of many parameters as previously mentioned in Section 4.4.4. Therefore, the less related the two species are, the more distinct their dissociation curves will be. Thus, the range of the equilibrium curves for water-breathing fish will always be larger than that for air-breathing ones simply because the water-breathing fish species are more numerous and more diversified.

In fact, when closely related species are compared, such as *Hoplias* and *Hoplerythrinus*, or *Arapaima* and *Osteoglossum*, a substantial difference exists: the blood of the water-breather has higher O_2 affinity (Johansen et al. 1978b; Galdames-Portus et al. 1979; Riggs et al. 1979). Similar results have been observed for the facultative air-breathing *Pterygoplichthys multiradiatus* when breathing in the water and air (Val et al. 1990). In this last example, there is no specific differences; the higher O_2 affinity during water-breathing is the consequence of physiological and biochemical adjustments.

4.4.5 The Root Effect

Root effect is a functional property of one or more hemoglobins in some teleost fish. This property consists of a reduction in the O_2 carrying capacity of the blood when the pH values are low, even if the O_2 tensions are high (Root 1931; Scholander and van Dam 1954). Differences in the expression of the Root effect, like the Bohr effect, can be related to the primary structures of isohemoglobins. This effect was considered as an exaggerated Bohr effect by Farmer et al. (1979). However, as pointed out by Britain (1987), normal hemoglobins over the pH range of 5–10, exhibit a Bohr effect, which consists of two parts. Below about pH 6.5, as the pH decreases, a slight increase in O_2 affinity is observed. This part of the curve is often referred to as the acid Bohr effect. As the pH increases above 6.5 a significant increase in O_2 affinity is observed and this part of the curve is called the alkaline Bohr effect or just, the Bohr effect. Thus, it seems erroneous to interpret the Root effect as an exaggerated Bohr effect because, at low pHs, their effects are in opposite directions.

It has been proposed that the presence of hemoglobins which exhibit a marked Root effect will facilitate the secretion of O_2 into the swim bladder and eye through the rete mirabile. It has been suggested that acidification occurs as the blood passes through the rate mirabile and, as a consequence, the O_2 carried by these hemoglobins is released locally, increasing the partial pressure of O_2, which then diffuses into the swim bladder or eye (Kuhn et al. 1963; Steen 1963). Based on structural similarities, Wittenberg and Wittenberg (1974) proposed that both retia function in similar ways. Ingermann and Terwilleger (1982), studying 15 species of marine teleosts, proposed that the presence of hemoglobins with a marked Root effect is much more related to the presence of choroid rete mirabile in the eye. These authors showed that fish which lack a swim bladder but still contain a choroid rete do have hemoglobins with a marked Root effect and fish lacking both swim bladder and choroid rete, such as some chondrichthyans, do not have a pronounced Root effect. *Lepidosiren paradoxa* has neither a swim bladder nor a choroid rete and, as expected, does not have any hemoglobins with a marked Root effect (Farmer et al. 1979; Peixoto and Val, unpubl. data).

While analyzing different fish species of the Amazon, we observed the highest magnitude of the Root effect in those species having both retia. The lowest values were detected in those species that do not have a choroid rete or a swim bladder. The species having only rete mirabile in the choroid produce dissociation values close to the values of the animals having both retia (Fig. 4.26). Farmer et al. (1979), also studying fishes from the Amazon, observed the highest values in animals having both retia. According to Farmer et al. (1979), a strong Root effect in animals which have a choroid rete but lack the swim bladder rete suggests that the choroid rete mirabile may be a more primitive structure and thus associated with the origin and evolution of hemoglobins with a marked Root effect.

No clear relationship has been observed between Root effect and number of Hb fractions. Nevertheless, Farmer et al. (1979) showed that the Hb component II of *Hoplosternum littorale* exhibits a significant Root effect while the Hb component I has a reverse Bohr effect which suggests that there is some compartmentalization of the functions of these hemoglobins. In addition, we failed to show any relationship between occurrence of

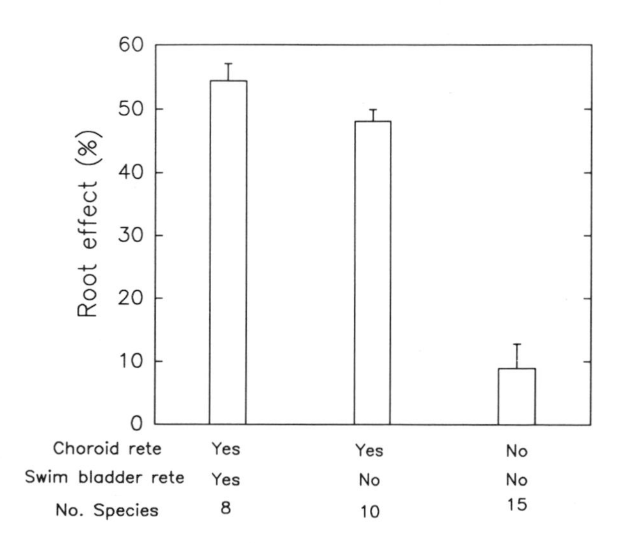

Fig. 4.26. Relationship between presence of rete mirabile and Root effect in fishes of the Amazon. (Data from Peixoto-Neto and Val, unpubl.)

hemoglobins with a marked Root effect and level of activity, O_2 resistance, trophic level, and habitat preferences of the fish species that were analyzed (Table 4.5). These results strongly support the hypothesis that there is a relationship between hemoglobins with the Root effect and the presence of rete mirabile.

The magnitude of the Root effect in several freshwater fish is increased by the addition of ATP and GTP (reviewed by Pelster and Weber 1991). In addition, it has been shown that GTP augments the Root effect more than ATP (Pelster and Weber 1990). This is also true for the Amazonian fish species which have thus far been analyzed. It is interesting, however, to observe that, despite the fact that the organic phosphates can induce a Root effect in some species (Pelster and Weber 1990), their effect seems to be related to the tendency of the Hb molecule to become desaturated at low pHs due to structural characteristics. In fact, the addition of NTP to the stripped hemoglobins, which display no Root effect, of some Amazonian fish species such as *Sorubim lima*, *Pinirampus pirinampu*, *Pterygoplichthys multiradiatus*, *Pimelodus blochii*, among other silurids, does not induce any decrease in the O_2 saturation levels. This clearly suggests that organic phosphates only affect those hemoglobins exhibiting the Root effect, even if its magnitude is zero for the stripped molecule. In other words, the organic phosphates do not induce the Root effect in all hemoglobins.

Because the occurrence of rete mirabile follows a phylogenetic pattern, it is not surprising that a relationship between the magnitude of the Root effect of hemoglobins and the phylogenetic status of the animal exists (Farmer et al. 1979; Peixoto and Val, unpubl. data). Despite the large variation, Characiformes, Gymnotiformes, and Siluriformes have, on average, desaturation values of 50, 30, and 5%, respectively, at low pHs (Root effect). Except *Apteronotus* sp. and *Hoplosternum littorale* (36 and 21% of Root effect, respectively), the Root effect is absent in most of the silurids that we analyzed in the Amazon. Some of them, however, exhibit hemoglobins with a very small Root effect, no more than 2% of desaturation. Similarly, general findings have been reported for other silurids of the Amazon and from other environmental conditions (Farmer et al. 1979; Dafré and Wilhelm 1989).

This general trend is also supported by the importance of the eyes for detecting food and predators. This ability is exceptional in Characiformes, intermediary in Gymnotiformes, and reduced in many Siluriformes. Nelson (1984), for example, pointed out that the eyes are usually small in Siluriformes and the barbels are important in detecting food. In addition, Ali and Klyne (1985) and Ali and Anctil (1976), who analyzed some silurids, showed that the cones in the eyes are generally small and only single cones have been observed in the poorly developed eyes. As pointed out before (Chap. 3), several species of catfishes are bottom-dwellers and many are nocturnal; this type of life-style reduces the importance of the eyes. The

Table 4.5. Root effect in hemolysates of selected fish species of the Amazon, with reference to their life-style, hypoxia tolerance, and trophic level

Species	Root effect (%)	Life-style	Hypoxia tolerance	Trophic level	Rete mirabile[a]	
					Swim bladder	Choroid
Cichla ocellaris	60	Active	No	Carnivore	Yes	Yes
Astronotus ocellatus	60	Moderate	Yes	Omnivore	Yes	Yes
Pellona castelnaena	57	Active	No	Carnivore	No	Yes
Piaractus brachypomum	53	Active	Yes	Herbivore	No	Yes
Plagioscion squamosissimus	50	Active	No	Carnivore	Yes	Yes
Serrasalmus elongatus	50	Active	No	Carnivore	No	Yes
Serrasalmus nattereri	49	Active	No	Carnivore	No	Yes
Arapaima gigas	47	Moderate	—	Carnivore	Yes	Yes
Schizodon fasciatus	47	Active	No	Herbivore	No	Yes
Colossoma macropomum	46	Active	Yes	Herbivore	No	Yes
Leporinus trifasciatus	44	Active	No	Omnivore	No	Yes
Rhytiodus microlepis	40	Active	No	Herbivore	No	Yes
Ageneiosus brevifilis	7	Active	No	Carnivore	No	No
Pimelodus blochii	5	Moderate	No	Omnivore	No	No
Sorubim lima	4	Moderate	No	Carnivore	No	No
Hypophthalmus marginatus	2	Active	No	Omnivore	No	No
Pinirampus pirinampu	1	Moderate	Unknown	Carnivore	No	No
Callophysus macropterus	1	Moderate	No	Omnivore	No	No
Hypophthalmus fimbriatus	0	Active	No	Omnivore	No	No
Pterodoras granulosus	0	Active	Unknown	Omnivore	No	No

[a] Information deducted from Farmer et al. (1979) and Peixoto-Neto and Val (unpubl.).

gymnotiforms are also nocturnal (Nelson 1984), using electroreception to interact with the environment (Kramer 1990).

4.4.6 Intraerythrocytic Phosphates

ATP and GTP are the major organic phosphates found in the red blood cells of the great majority of fishes, including the Amazonian ones which have been analyzed (Bartlett 1980; Nikinmaa 1990; Val 1993). A few species have other types of organic phosphates in their red blood cells, including 2,3 diphosphoglycerate (2,3-DPG), inositol diphosphate (IP_2), and inositol pentaphosphate (IPP). As mentioned in Section 4.4.5, these organic phosphates strongly reduce the O_2 affinity of a variety of fish hemoglobins. In general, the strength of the effect that organic phosphates have on O_2 affinity of teleost hemoglobins decreases in the following order IHP>GTP>ATP> 2,3-DPG. However, some researchers have found that IPP rather than IHP (inositol hexaphosphate) is present in fish red blood cells and that ATP and GTP have equal effects on some fish hemoglobins (Weber and Lykkeboe 1978; Gronenborn et al. 1984; pers. observ.). Intraerythrocytic concentrations of organic phosphates in fish show a wide intra- and interspecific variation (Table 4.6). The intraspecific variation in the levels of ATP and GTP has been related to many physiological and environmental factors such as exercise, temperature, nutrition, ontogeny, and O_2 supply (Bartlett 1980; see Nikinmaa 1990).

4.4.6.1 Ontogenetic Changes in the Red Blood Cell Organic Phosphate Concentrations

Inositol pentaphosphate seems to be a distinguishing feature of *Arapaima gigas*. It was first described in the erythrocytes of *Arapaima gigas* simultaneously by Isaacks et al. (1977) and Bartlett (1978b). Previously, IPP was known to occur only in the erythrocytes of birds and turtles and nowhere else in nature. Johnson and Tate (1969), studying this compound from red cells of birds, proposed that it is myo-inositol 1,3,4,5,6-pentaphosphate. Although myo-inositol is a common metabolite, its phosphorylation seems to be a specific feature of the erythrocytes of the above-mentioned groups of animals. In contrast to ATP and GTP, IPP is not a common metabolic intermediate of oxidative phosphorylation, nor can its intraerythrocytic concentration be adjusted by a change in the regulatory properties of a single enzyme. In birds and turtles a qualitative and quantitative ontogenetic change of the red blood cell's organic phosphates, including IPP, has been reported (Bartlett and Borgese 1976; Bartlett 1978c).

In common with birds and turtles, the intraerythrocytic concentration of IPP in *Arapaima* shows an ontogenetic variation (Val et al. 1992b; see Fig. 4.27). In birds, it appears first in the late embryo, and within a few days

Table 4.6. The major intraerythrocytic phosphates of selected fish species of the Amazon, expressed as mM/l of red blood cells. The orders are organized from the generalized to the specialized according to Nelson (1984)

Families and species	2,3- DPG	ATP	GTP	IPP	Source
Rajiformes					
Potamotrygon motoro	–	0.4	0.1	–	Johansen et al. (1978a)
Lepidosireniformes					
Lepidosiren paradoxa	–	1.6	3.1	–	Isaacks et al. (1978)
Lepidosiren paradoxa	–	1.4	0.6	–	Johansen et al. (1978a)
Osteoglossiformes					
Arapaima gigas	–	2.1	2.6	3.1	Isaacks et al. (1978)
Arapaima gigas (length = 17.30 cm)	–	2.6	1.7	2.3	Val et al. (1992b)
Arapaima gigas (length = 107 cm)	–	1.6	0.7	6.5	Val et al. (1992b)
Osteoglossum bicirrhosum	–	0.4	0.3	–	Johansen et al. (1978a)
Osteoglossum bicirrhosum	–	7.1	5.2	–	Bartlett (1978b)
Characiformes					
Hoplias malabaricus	–	0.4	0.3	–	Johansen et al. (1978a)
Hoplerythrinus unitaeniatus	–	0.9	0.4	–	Johansen et al. (1978a)
Prochilodus cf. *nigricans* (July)	–	1.1	0.2	–	Val et al. (1992a)
Prochilodus cf. *nigricans* (November)	–	0.8	1.4	–	Val et al. (1992a)
Colossoma macropomum (January, lake)	–	2.4	0.5	–	Val (1986)
Colossoma macropomum (November, river)	–	1.4	2.4	–	Val (1986)
Piaractus brachypomum (January, lake)	–	2.9	0.6	–	Val (unpubl.)
Semaprochilodus insignis	–	0.3	0.9	–	Val et al. (1984)
Semaprochilodus taeniurus	–	0.3	0.5	–	Val et al. (1984)
Brycon cf. *cephalus*	–	1.0	–	–	Val and Almeida-Val (1988)
Brycon cf. *erythropterum*	–	1.6	1.5	–	Val and Almeida-Val (1988)

Table 4.6. *Continued*

Families and species	2,3- DPG	ATP	GTP	IPP	Source
Siluriformes					
Hoplosternum littorale	–	0.8	1.6	–	Johansen et al. (1978a)
Hoplosternum littorale (March)	1.1	2.6	8.4	–	Affonso (1990)
Hoplosternum littorale (June)	0.6	0.5	1.2	–	Affonso (1990)
Dianema urostriatum	–	2.1	8.3	–	Marcon (unpubl.)
Pterygoplichthys multiradiatus[a] (wild)	–	2.5	0.5	–	Val et al. (1990)
Pterygoplichthys multiradiatus[a] (normoxia)	–	3.4	4.8	–	Val et al. (1990)
Pterygoplichthys sp.	–	1.4	1.1	–	Johansen et al. (1978a)
Callophysus macropterus (white water)	–	2.0	1.3	–	Ramirez-Gil (1993)
Callophysus macropterus (black water)	–	2.5	2.8	–	Ramirez-Gil (1993)
Pseudoplatystoma sp.	–	1.6	3.4	–	Isaacks et al. (1978)
Gymnotiformes					
Electrophorus electricus	–	0.6	0.3	–	Johansen et al. (1978a)
Synbranchiformes					
Synbranchus marmoratus	–	1.1	0.9	–	Johansen et al. (1978a)
Perciformes					
Plagioscion squamosissimus	–	0.4	0.2	–	Val (unpubl.)

[a] *Liposarcus pardalis*.

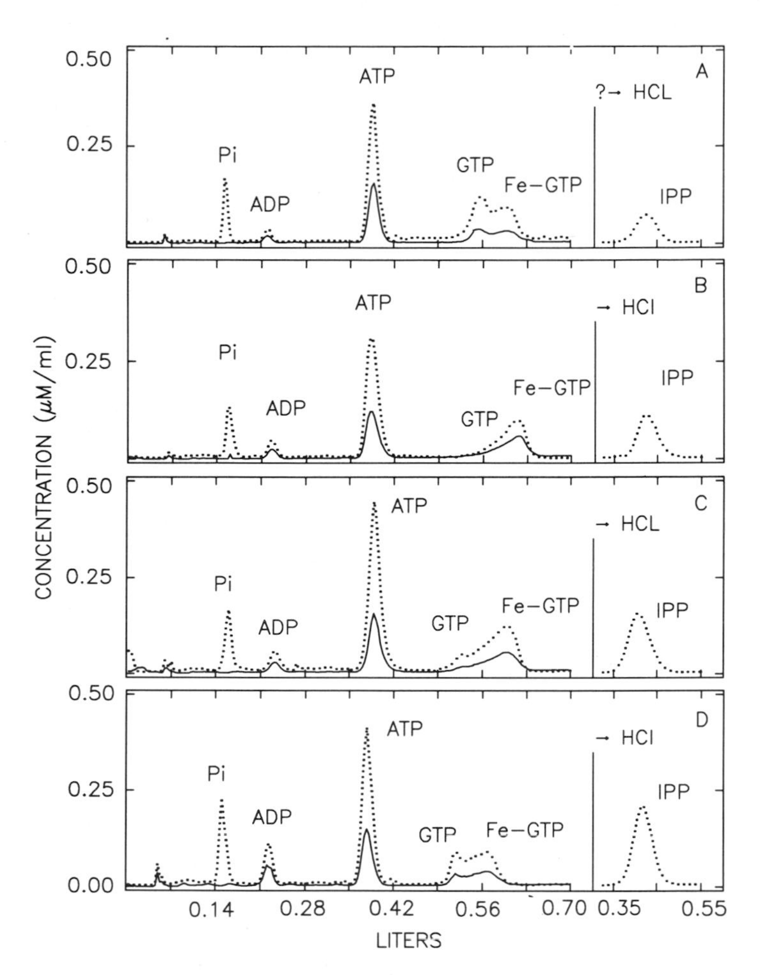

Fig. 4.27. Ion chromatographic profile of intraerythrocytic phosphates of *Arapaima gigas* at different ages (lengths). Observe the significant increase in IP_5 levels from **A** to **D**. Quantities are expressed in μM/ml of eluate. *Solid line* shows optical density at 260 nm calculated as micromoles of adenine and *dotted line* shows total phosphorus (μM). (Val et al. 1992b)

after hatching, its concentration increases rapidly reaching the characteristic high concentration of adults (Bartlett 1980; Isaacks and Harkness 1980). In turtles and *Arapaima*, on the other hand, its concentration increases very slowly stabilizing only 1 year after its appearance. In addition, as the IPP level increases, the ATP and GTP levels decrease. Because no significant changes in hematocrit and Hb concentration is observed during the first year of life of *Arapaima* (Salvo-Souza 1990), the result is a significant increase in the IPP:Hb ratio simultaneously with a slight decrease in the NTP:Hb ratios (Fig. 4.28).

Juvenile *Arapaima* are normal water-breathers, therefore, the slow increase in the concentration of the strongest $Hb-O_2$ modulator (IPP) is

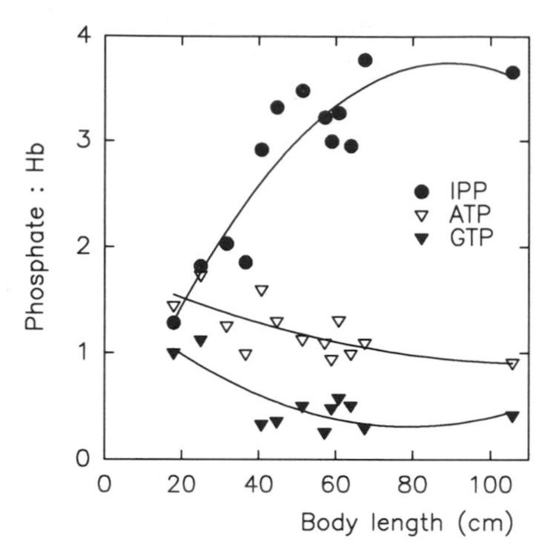

Fig. 4.28. Intraerythrocytic phosphate: Hb ratios during the development of *Arapaima gigas*. Note the significant increase in IPP:Hb ratio, followed by a slight decrease in ATP:Hb and GTP:Hb ratios. (Val et al. 1992b)

advantageous during the first year of life. If high levels of IPP are present in the erythrocytes of the young animals, the O_2 uptake at the gills will be impaired. However, the presence of a strong allosteric cofactor of $Hb-O_2$ is far more advantageous for O_2 delivery to tissues in the adults.

Except for *Arapaima gigas*, no ontogenetic changes in the concentration of intraerythrocytic organic phosphates have been observed in the Amazonian fish thus far. In fact, ontogenetic changes in the concentration of NTP such as those which occur in the viviparous fish *Embiotoca lateralis* (Ingermann and Terwilleger 1981) seem to be an unusual feature of fish.

4.4.6.2 Effects of Migration on Intraerythrocytic NTP Levels

The main intraerythrocytic phosphates detected in the species of the genus *Semaprochilodus*, including a possible hybrid, are ATP and GTP (Val et al. 1984, 1986). The specimens of this genus migrate twice a year, but for some unknown reason, part of the school does not migrate. The nonmigrating fish have about ten times less ATP and two times less GTP than their migrating counterparts. As a consequence, the magnitude of the whole blood Bohr effect is smaller in the nonmigrating animals.

4.4.6.3 Environmental Factors Affecting the Red Blood Cell Organic Phosphate Concentrations

Dissolved O_2 has been described as the main factor influencing the intraerythrocytic level of organic phosphates in fishes. However, other environmental factors influence the concentration of these allosteric Hb effectors, including temperature and water level oscillations. In general, as these environmental factors oscillate beyond the optimum for the animal and thus creating a stressful situation, a decrease in the concentration of intraerythrocytic organic phosphates is observed. The changes in NTP levels have always been related to changes in the O_2 uptake capacity from the environment.

4.4.6.3.1 Effects of Oxygen Availability

A significant decrease in the concentration of ATP and GTP in fish exposed to hypoxia has been described for many species, including several from the Amazon basin (Bartlett 1980; Soivio et al. 1980; Tetens and Lykkeboe 1981; Val et al. 1990; Val 1993). As in other fish species, the concentration of GTP varies faster than the concentration of ATP in Amazon fishes which are exposed to natural oscillations in the amount of available O_2. Greaney and Powers (1978) suggested that the decrease in the concentration of ATP in the red blood cells of fish that have been exposed to hypoxia is a direct consequence of an inadequate O_2 supply to the oxidative phosphorylation processes within the fish. However, this explanation has not been accepted because the ATP levels in many other cell types are affected only when the O_2 tension decreases below 5 mm Hg (reviewed by Nikinmaa 1990). Recent studies have suggested that the intraerythrocytic concentration of ATP can be controlled by catecholamines released during low O_2 tension exposures (Tetens et al. 1988). However, ATP levels change even during small changes in the O_2 availability, apparently before the catecholamines have been release into the circulation (Val 1986; Val et al. 1994). In addition, in vitro experiments with carp red blood cells have indicated that the catecholamines do not affect the GTP concentration (Salama and Nikinmaa 1988).

Considering the interconversion reactions of ATP and GTP (Fig. 4.29) and their different modulatory effects, it is possible that the ATP levels are maintained during the initial phase of hypoxia as a result of GTP degradation. The formation of complexes of NTP with metal ions may also control their intraerythrocytic levels. GTP and especially ATP are readily bonded with magnesium. This greatly reduces their modulatory effect on $Hb-O_2$ affinity (see Houston 1985). GTP, on the other hand, is readily complexed with Fe, forming Fe-GTP (Bartlett 1976). Fe-GTP was first detected in rat reticulocytes and it was suggested that the formation of the complex occurs as a result of the reaction of GTP and Fe during the extraction with perchloric acid (Bartlett 1976). This complex has been detected in the red blood cells of many fish species (Bartlett 1980; Monteiro et al. 1987; Val et al. 1990). Using the same extraction method, we have shown large variations in the intraerythrocytic concentration of Fe-GTP in some Amazonian fishes which is sometimes related to dissolved O_2 levels. No correlation has been observed between Fe-GTP and concentrations of Hb and/or transferrins. Interestingly, there is a tendency for the formation of a positive, linear relationship between intraerythrocytic concentrations of

Fe-GTP and GTP (Fig. 4.29). The physiological role of Fe-GTP is unknown but it is possible that it has a reduced modulatory effect on $Hb-O_2$ affinity similar to Mg-ATP or Mg-GTP.

The intraerythrocytic concentration of ATP and GTP decreases whenever the animal is exposed to hypoxia. This is also valid for some facultative air-breathers. *Pterygoplichthys multiradiatus*, for example, who was studied at three different environmental conditions, exhibited a significant decrease in the concentration of both ATP and GTP when exposed to hypoxia, despite its capacity to breathe air (Fig. 4.30). Interestingly, the concentration of ATP in animals acclimated for 30 days to normoxia or hypoxia is about 1.4 times higher than in the animals exposed to natural hypoxic conditions. However, the concentration of GTP in animals acclimated to normoxic conditions is about eight times higher than in those exposed to hypoxia (acclimated or under natural hypoxia). These results suggest that hypoxia, under natural conditions, is more stressful than under controlled conditions in the laboratory.

To be effective, the adaptive changes in the concentration of intraerythrocytic phosphates should be rapid enough to "buffer" the diel oscillations of dissolved O_2 levels. For example, Tetens and Lykkeboe (1981) reported a significant reduction in red cell NTP concentration within 1 h when the trout that they were studying were exposed to deep hypoxia. Ferguson and

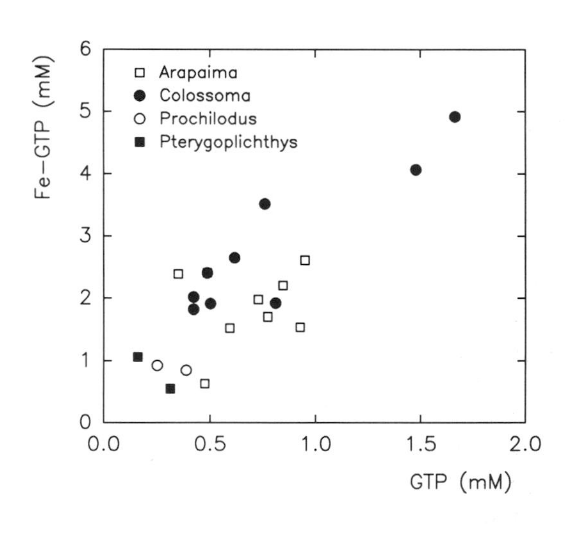

Fig. 4.29. Relationship between GTP and Fe-GTP in the red blood cells of selected fish species of the Amazon. The chemical equation depicts possible interconversion of ATP and GTP. (Val 1993)

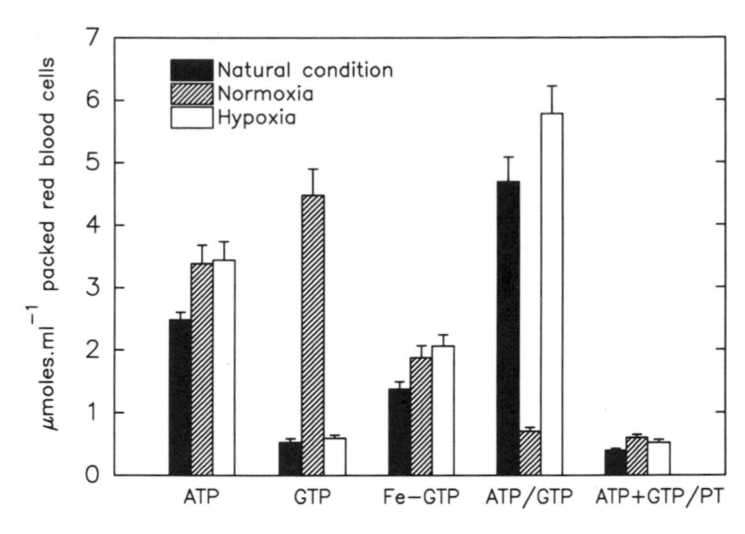

Fig. 4.30. Amounts of the main intraerythrocytic phosphates of *Pterygoplichthys multiradiatus* (= *Liposarcus pardalis*) under natural conditions, exposed to hypoxia and to normoxia. Quantities are expressed in μM/ml of eluate. (Val, unpubl.)

Boutilier (1989) suggested that some adaptive modification may occur at the level of the red cell membrane that induces a promptly regulation of NTP levels. Diel changes in the concentration of NTP inside the red blood cells have been observed for several fish species from the Amazon. For some of them, there is a direct correlation between these changes and the concentrations of dissolved O_2; for others no apparent correlation is observed; and finally, for others an inverse correlation is observed as exemplified by *Prochilodus* cf. *nigricans* (Val et al. 1992a). Thus, under natural conditions and for some species, some other physiological or even environmental factors overrides the expected effect of O_2 availability on intraerythrocytic levels of NTP.

The seasonal variations in the concentrations of intraerythrocytic phosphates are very often weakly related to the amount of dissolved O_2 for fishes of the Amazon. They have been primarily related to seasonal changes in the river water levels and temperature. Nonetheless, a significant relationship between the intraerythrocytic levels of NTP and the Bohr effect of whole blood is invariably observed (Fig. 4.31).

4.4.6.3.2 Effects of Temperature

An increase in the environmental temperature appears to induce a decrease in the concentration of intraerythrocytic organic phosphates in fish (Nikinmaa et al. 1980; Houston and Koss 1984). This general trend has been observed even during small changes in the temperature. Nikinmaa et al. (1980), for

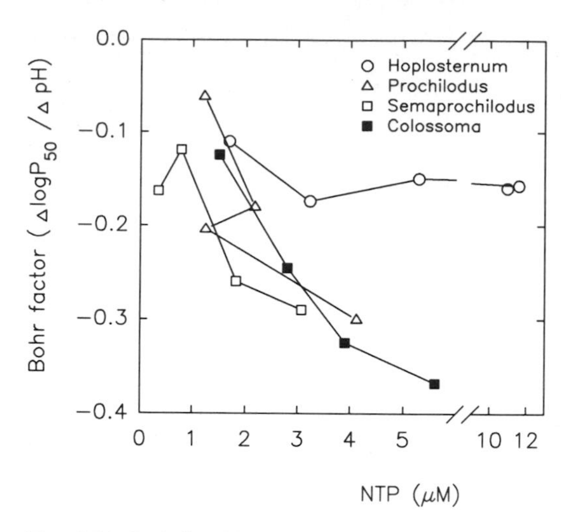

Fig. 4.31. Relationship between intraerythrocytic levels of NTP (μM) and whole blood Bohr effect of selected fish species of the Amazon. (Val, unpubl.)

example, observed a significant decrease in ATP concentrations in rainbow trout when the temperature was increased from 15 to 18 °C. One exception to this pattern is the elasmobranch, *Cephaloscyllium isabella* (Tetens and Wells 1984). Intraerythrocytic levels of organic phosphates in the fishes of the Amazon are also affected by temperature (both controlled changes in the laboratory and natural changes).

Besides ATP and small amounts of GTP, a significant concentration of 2,3-DPG has been detected in the erythrocytes of *Hoplosternum littorale* (Fig. 4.32). This latter organic phosphate is not a common phosphate in the red blood cells of fish (Bartlett 1980). Erythrocytic 2,3-DPG was detected only in *Cichlasoma cyanoguttatun* (Gillen and Riggs 1971) and in *Pterygoplichthys* spp. (Isaacks et al. 1978). Studying *Pterygoplichthys multiradiatus* (possibly the same species studied by Isaacks and coworkers), Val et al. (1990) failed to detect any 2,3-DPG. Thus, as far as we know, 2,3-DPG is restricted to the red cells of two or three species of fish, all of them occurring in the Amazon.

The intraerythrocytic concentrations of 2,3-DPG in *Hoplosternum* exhibited significant changes which were correlated with the thermal regime of their habitat in the Amazon, i.e., the higher the temperature, the higher the concentration of 2,3-DPG inside the red blood cells (Fig. 4.33A). In addition, Affonso (1990) observed a direct relationship between the concentration of 2,3-DPG and the magnitude of the Bohr effect of whole blood, and consequently, between the Bohr effect and the temperature of the animal's habitat (Fig. 4.33B). In general, as the environmental temperature increases, the amount of dissolved O_2 in the water decreases. This causes the specimens of *Hoplosternum* to surface in order to breathe air. As the

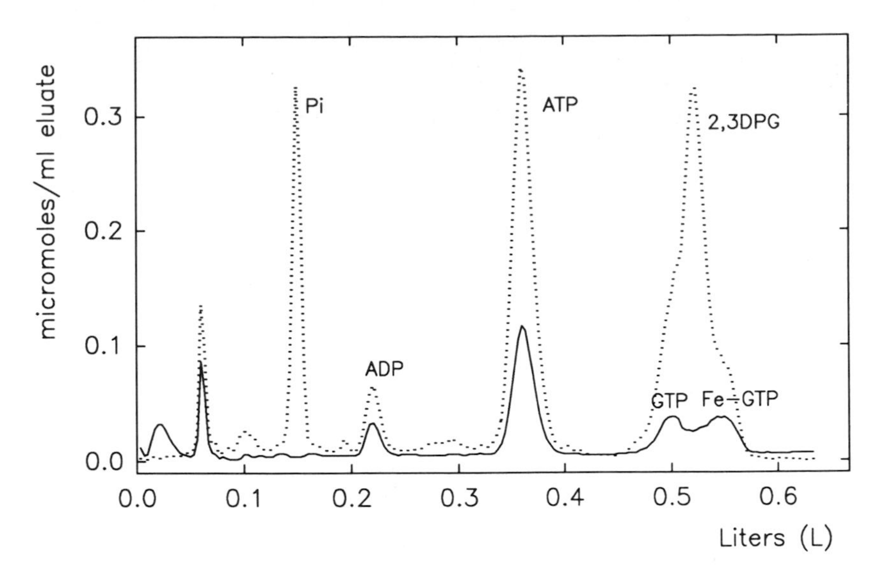

Fig. 4.32. Ion chromatographic profile of intraerythrocytic phosphates of *Hoplosternum littorale*. Quantities are expressed in μM/ml of eluate. *Solid line* shows optical density at 260 nm calculated as micromoles of adenine and *dotted line* shows total phosphorus (μM). (Affonso et al., unpubl.)

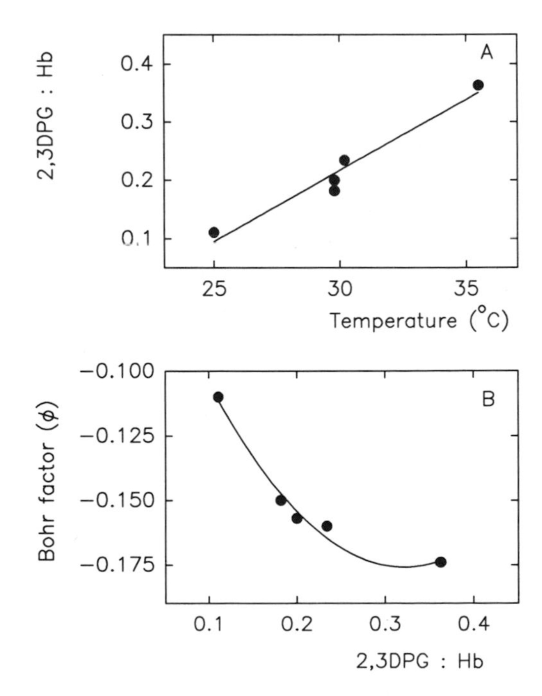

Fig. 4.33. Relationships between **A** environmental temperature and intraerythrocytic 2,3-DPG and **B** 2,3-DPG and whole blood Bohr factor in *Hoplosternum littorale*. (Affonso et al., unpubl.)

animal relies on air-breathing, an increase in buffering capacity of the blood is observed, thus, an increase in the magnitude of the whole blood Bohr effect is advantageous. No correlation was observed between 2,3-DPG levels and dissolved O_2 concentrations (Affonso 1990).

4.4.6.3.3 The Effects of Oscillations in the Water Levels of Rivers in the Amazon

Water level oscillations in the Amazon rivers, as mentioned in Chapter 2, are an important driving force of the biota in the Amazonian ecosystems. It does not represent per se an environmental pressure. The water levels, however, can be seen as indices of the interactions of many different biotic and abiotic factors. Thus, it is not surprising to detect a correlation between a physiological and/or biochemical parameter and the elevation of the river. In fact, in many fish species of the Amazon, correlations between river level and activities of some respiratory enzymes, hematocrit, blood concentration of Hb, concentration of specific Hb fractions, and levels of intraerythrocytic phosphates, among others, have been found (Val 1983, 1986, 1993; Almeida-Val 1986; Monteiro et al. 1987; Almeida-Val and Val 1990; Almeida-Val et al. 1991b; Val et al. 1992a).

Monteiro et al. (1987), studying *Mylossoma duriventris*, showed that, as the water level increased in the river, the intraerythocytic levels of ATP and GTP decreased (Fig. 4.34A). In this study, the authors showed a slight inverse correlation between the concentration of dissolved O_2 at 50-cm depth and the river water level oscillation. As a consequence, the expected correlation between NTP (particularly GTP) and concentrations of dissolved O_2 was not observed for this fish species under natural environmental conditions. Val et al. (1992a) also failed to show the expected correlation between intraerythrocytic levels of NTP and dissolved O_2 levels for *Prochilodus* cf. *nigricans* when exposed to natural conditions. However, as in *Mylossoma*, the intraerythrocytic levels of NTP was higher when the river water level was low. Similar results have been reported for *Colossoma macropomum* and species of the genus *Semaprochilodus* (Val 1983, 1986).

4.4.6.4 Interactive Effects

Many of the adjustments to improve the O_2 transfer in fishes of the Amazon (in their natural environment) occur simultaneously, as if they were induced by a main factor. These adaptations include changes in the relative concentration of isohemoglobins; changes in the intraerythrocytic levels of ATP, GTP, and 2,3-DPG; changes in red blood cell numbers, blood Hb concentration, and hematocrit; adjustments of the activities of some respiratory isozymes; morphological modification of the lower lip; and gill hyperventilation. In *Colossoma macropomum*, for example, as the intraerythrocytic concentration of NTP decreases, the hematocrit increases (Val

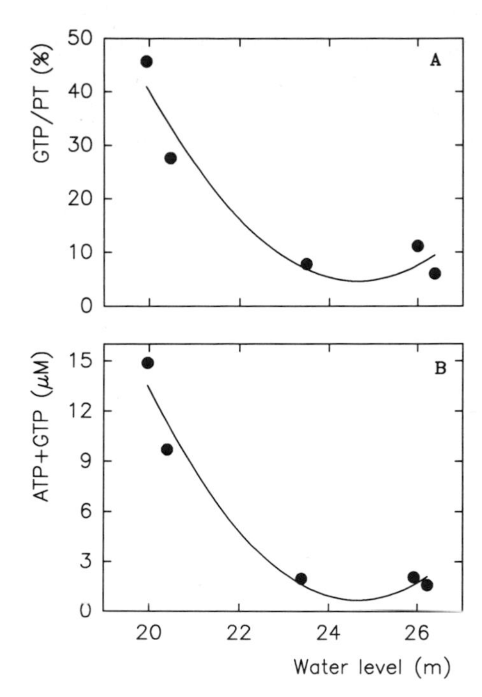

Fig. 4.34. Relationships between water level and **A** red cell GTP/PT and **B** ATP+GTP in *Mylossoma duriventris*. (Monteiro et al. 1987)

1993). The studies which were carried out over the past few years have revealed that the catecholamines released into the blood circulation of fish species from climatic zones other than the Amazon, play an important role in almost all processes related to O_2 transfer in fishes (Randall 1990). A wide variety of environmental and physiological conditions, including hypoxia, hypercapnia, anaemia, exercise, blood acidosis, and air exposure, induce the release of catecholamines in these fish species (Randall 1990; reviewed by Randall and Perry 1992). The increase in concentration of circulating catecholamines has been correlated with many physiological and biochemical changes which are directed towards enhancing O_2 transfer (Nilsson and Grove 1975; Wood 1976; Isaia et al. 1978; Primmett et al. 1986; Milligan and Wood 1987; Walsh et al. 1988; Perry and Kinkead 1989; Nikinmaa 1990). Thus, it is quite possible that the "main factor" (as alluded to above), which induces the simultaneous adaptive adjustments observed in fish of the Amazon under natural conditions, is the release of catecholamines.

4.5 Effects of Specific Environmental Conditions on Gas Transfer

Distinctive environmental changes, for example, sudden increases in the concentration of dissolved H_2S, particularly during the *friagem* period, sudden releases of anoxic water downstream by hydroelectric power plants, and petroleum contamination will often kill many fish in the Amazon. The mortality under these circumstances occurs due to two main factors: (1) the inability of the animal to anticipate the sudden environmental change and to buffer their effects, and (2) the transient or even permanent damage caused by the contaminant. For example, it has been shown that the Hb reacts with H_2S reducing its capacity to transport O_2 (Jones 1964). Affonso (1990) suggested that specimens of *Hoplosternum littorale* survive high H_2S concentration in their natural environment mainly due to their ability to keep breathing air even at high levels of dissolved O_2 in the water.

Chapter 5

Respiratory Metabolism

Due to the environmental characteristics of the Amazon rivers, the organisms have only one choice: they must survive chronic hypoxia. High glycolytic capacities are always expected in organisms that survive long periods of anoxia, and it happens to occur in anaerobes such as the aquatic turtle and the goldfish (Hochachka 1980). On the other hand, more recent developments in this field have shown that surviving chronic O_2 limitations involves different metabolic changes in the organisms. These changes include (1) lowering the oxidative capacity in order to lower the demand for O_2 and (2) maximizing O_2 utilization, i.e., the production of more ATP per mole of O_2 consumed (Hochachka 1993).

The Amazon ichthyofauna faces two kinds of challenges regarding O_2 availability in the water: first, it has to cope with chronic hypoxia (low O_2 contents in the water during the whole life span), and second, it may confront anoxia periodically (on some occasions, almost daily). The periods of anoxia are short when compared to those occurring in temperate regions.

Currently, there are four main ideas involved in the studies of the main defense strategies of the organisms against hypoxia: (1) preferential utilization of advantageous metabolic pathways, maximizing the ATP yield per mole of fermented substrate or consumed O_2; (2) down-regulation of metabolism, with a decrease in the rate of the processes which utilize ATP and the optimization of those which produce ATP; (3) preservation of integrated membrane functions, despite low ATP turnover rates; and (4) conservation of appropriate conditions to activate biochemical processes once the O_2 is available again (Hochachka and Somero 1984). Moreover, current studies have demonstrated that these four mechanisms exist in a coordinated way in several vertebrate and invertebrate groups (reviewed by Hochachka and Guppy 1987).

The first studies on respiratory metabolism in fishes of the Amazon occurred during the expedition of the Research Vessel Alpha-Helix in 1976. After studying the enzyme profiles and ultrastructure of different tissues from air-versus water-breathing fish, one of the main conclusions was that there is preferential reliance upon oxidative metabolism in both groups, besides the capacity of balancing anaerobic and aerobic pathways (reviewed by Almeida-Val et al. 1993). Since then, there has been very little development in this field, apart from some studies on isozymes and their functional characteristics according to seasonal fluctuations in environmental parameters.

By comparing the metabolic rates of similar organisms with different types of respiratory patterns we can acquire a good indication of the animal's needs and the way it uses O_2. Scholander et al. (1953) were the first to report the metabolic rates of tropical fish who were exposed to different temperatures and compared this data to that for arctic fish. Following this study, little new information has been reported, leaving an enormous lacuna in the study of this area in Amazonian fishes. We will summarize the available data for Amazon fish, comparing air- versus water-breathing fish and describing the reported patterns of responses to hypoxia.

5.1 Optimizing Aerobic Metabolism

Metabolic rates of ectothermic vertebrates are always an order of magnitude lower than endotherms of a similar size (Dejours 1979; Else and Hulbert 1985). From studies of both groups, it is well known that O_2 uptake (based on weight unit) varies inversely with animal size and may also be affected by temperature changes.

As mentioned above, the first measurements of O_2 uptake in tropical fishes were reported by Scholander et al. (1953) from studies comparing temperature adaptation of arctic and tropical poikilotherms (vertebrates and invertebrates). Particularly in fish, higher metabolic rates were reported in the tropical species than in the arctic fish species. Furthermore, if the arctic fish happen to face tropical temperatures, their O_2 intake would be much higher than the tropical ones and vice versa; this indicates that both groups are perfectly adapted to their own thermal niches (Scholander et al. 1953). Very little has been done in terms of measurements of metabolic rates of tropical fish since that report, with the exception of a few studies on lungfish by Kjell Johansen and his colleages (Johansen et al. 1967, 1970; Johansen and Lenfant 1967, 1968) and on Indian air-breathing fish (Singh 1976). Indeed, the main interest has always been air-breathing fish species. Consequently, metabolic rates of water-breathing fish of the Amazon are almost unknown.

5.1.1 Air-Breathers Versus Water-Breathers

The main legacies of air-breathing in vertebrates are endothermy and high energy fluxes. The latter allows the former, i.e., it keeps the body warmed at 37 °C which helps maintain high metabolic rates with optimum energy turnover rates. This is the reason why mammals and birds (warm-blooded animals) exhibit much higher metabolic rates (O_2 consumption) than reptiles, amphibians, and fish (cold-blooded animals). Comparisons between similarly sized reptiles and mammals indicate lower metabolic rates and a smaller surface area of mitochondrial membrane in the reptiles (Else and Hulbert 1985). High energy turnover rates are required for almost all biological activities. However, endothermy and high energy fluxes are characteristically found only in a few water-breathing instead of air-breathing fish species (reviewed by Almeida-Val and Hochachka, in press). Tuna is the main example of endothermy in fish and retains one of the highest energy turnover rates (Hochachka 1979; Bushnell and Jones 1992). Air-breathing fishes, on the other hand, have very low energy turnover rates and low metabolic rates when compared to water-breathers (see below). If selection for air-breathing was not accomplished with energy gains, what were the selective forces that led to its evolution in fish? Given that this attribute is uncommon, it is interesting to consider why evolution of air-breathing occurred in fish.

One of the main propositions is that environmental constraints, mainly O_2 depletion, forced fish to search for O_2 at the water's surface and, accidentally, fish started to gulp air bubbles (Gans 1970). From morphological, physiological, and metabolic points of view, this solution should involve many complex mechanisms. In the next sections, air-breathing fish will be compared to their close water-breathing relatives, addressing the evolution of energy metabolism.

By comparing air- and water-breathing fishes it is possible to realize how successful both are in view of environmental O_2 depletion. In addition, it is possible to outline how differently they supply O_2 to the energy-dependent biological activities, such as swimming during predation or migration. It is well known that O_2-dependent processes generate much more energy than the fermentative pathways. When comparing metabolic rates of air- versus water-breathing species, one should be able to determine which is the more efficient type of respiration, i.e., which type produces higher oxidative metabolism with a consequently higher ATP yield. The very warm Amazonian climate, in itself, may be enough to increase metabolic activity as well as O_2 consumption. However, the available data show a different picture.

5.1.2 Oxygen Consumption

Of the tropical air- and water-breathing species listed in Table 5.1, the lowest metabolic rates occur in the African and in the South American (Amazonian) lungfish, followed by the Amazonian *Electrophorus*. Among the air-breathers, *Arapaima gigas* (pirarucu) has the highest O_2 consumption rate. Surprisingly, the non-Amazonian air-breathers and the temperate water-breathing fish present similar O_2 consumption rates which lie between Amazonian air-breathers and tuna, the most active fish.

Table 5.1. Total O_2 uptake ($\dot{V}O_2$ = $mg\,kg^{-1}\,h^{-1}$). Other species have been included for comparison

Species	$\dot{V}O_2$	O_2 from air (%)	References
Protopterus	11.40	91	McMahon (1970)
Lepidosiren[a] (adult)	22.20	96	Lenfant et al. (1970)
Electrophorus[a]	30.00	77	Farber and Rahn (1970)
Lepisosteus	53.40	73	Rahn et al. (1971)
Heteropneustes	84.55	41	Hughes and Singh (1971)
Clarias	93.90	58	Singh and Hughes (1971)
Amia	114.00	74	Johansen et al. (1970)
Arapaima[a]	129.00	82	Stevens and Holeton (1978b)
Hoplerythrinus[a]	174.00	40	Stevens and Holeton (1978a)
Onchorhynchus (trout)	114.32	0	Bushnell and Jones (1992)
Katsuwonus (tuna)	1177.50	0	Bushnell and Jones (1992)
Thunnus (tuna)	680.2	0	Bushnell and Jones (1992)
Juveniles			
Lepidosiren[a]	84.60	36	Johansen and Lenfant (1967)
Serrasalmus[a]	540.00	0	Almeida-Val (unpubl.)
Mylossoma[a]	420.00	0	Almeida-Val (unpubl.)
Triportheus[a]			
Rest	163.00	0	Holeton and Stevens (1978)
Active	433.00	0	Holeton and Stevens (1978)

[a] Selected water- and air-breathing fishes of the Amazon.

It is well known that mass-specific metabolic rates (i.e., O_2 consumption per unit of weight) vary inversely with the average size of the species in both endotherms and ectotherms. Because the pirarucu is the largest freshwater fish, its relatively high metabolic rate among the air-breathers represents a paradox. However, according to Schmidt-Nielsen (1984), these relationships for fish influenced by other parameters such as environmental temperature, thermal history of the species, O_2 content of the water, light-dark cycles, among others. In addition, the high metabolic rates observed for *Arapaima* in this study (Stevens and Holeton 1978a) could be related to the size of the animals analyzed. They were very small (1.87 to 3.04 kg) compared to the adult specimens whose weight can reach 250 kg (Salvo-Souza and Val 1990). Studying the lungfish *Lepidosiren* during different stages of its life, Johansen et al. (1970) observed that the juvenile specimens had higher specific metabolic rates than the adults and extracted only 36% of the inspired O_2 from the air. Furthermore, juveniles of *Arapaima* rely almost exclusively on dissolved O_2 during the first months of their life, increasing the frequency of air-breathing during the first year from approximately 10 to 90% (see Chap. 4). Nevertheless, pirarucu is large and sluggish and is considered to be a nonactive fish. Thus, it is quite possible that the metabolic rate of larger specimens of *Arapaima* is lower than those reported by Stevens and Holeton (1978a).

Another interesting fact is that metabolic rates in air-breathing fish are not exceptionally higher than those of water-breathers, nor even much higher than those of temperate fish. The development of the air-breathing behaviour in tropical fish should allow the organism to have a higher oxidative metabolism (higher intake levels of O_2) both by increasing the amount of available O_2 and by the effect of the higher environmental temperature. However, somehow, these fish live at the lower limits of energy turnover.

In general, facultative air-breathing fish are able to take in high levels of O_2, similar to those observed for temperate water-breathers (Table 5.1). Among them, the Amazonian jejú (*Hoplerythrinus unitaeniatus*) takes in one of the highest O_2 levels which reflects its high levels of activity (Stevens and Holeton 1978b; see Chap. 4). The metabolic rates of exclusively water-breathing fishes in the Amazon vary according to their activity levels. Comparisons between Amazonian and temperate fish species (Table 5.1) suggest that the amount of O_2 consumed per unit of weight is directly related to the activity of the fish. The best example is *Triportheus angulatus*, which has been evaluated during active and resting periods. Oxygen uptake is almost one order of magnitude higher in juveniles of water-breathing fish such as *Serrasalmus* (*Pygocentrus*) *nattereri* and *Mylossoma sp.* than in juveniles of the air-breathing *Lepidosiren*. *Serrasalmus* (*P.*) *nattereri* and *Mylossoma sp.* belong to the same group (Serrasalmidae) and are considered to be active fish with predatory and migratory habits, respectively. This explains their high metabolic rates. However, their intake levels of O_2 are much lower than those values obtained for tuna, which consume the greatest amount of O_2 of all fish (Bushnell and Jones 1992). Most Amazon fishes that demonstrate migratory behaviour are exclusively water-breathers and we assume that they take in O_2 at higher rates.

Thus, breathing air was not developed in fish with the aim of a more efficient energy turnover and, consequently, not for the purpose of better endurance. This last objective has already been accomplished in the water-breathing fish, such as tuna. It is most probable that air-breathing habits in fish were developed only to avoid the constraints imposed by low O_2 concentrations in the aquatic environment. The relatively high dependence of other metabolic activities, such as CO_2 excretion in water (see Chap. 4) supports this idea.

5.1.3 Enzyme-Level Adjustments

Two major types of enzyme-level adjustments are observed in air-breathing fish:

1. a down-regulation of absolute activities of enzymes of aerobic and anaerobic pathways; and
2. an up-regulation of ratios of glycolytic/oxidative capacities on a tissue-by-tissue basis (reviewed by Almeida-Val and Hochachka, in press).

Within this framework, the white muscles of fish can illustrate the main accomplishment of both patterns. For example, the activity of the enzyme lactate dehydrogenase (LDH) is a good measurement of anaerobic pathways, because it converts pyruvate to lactate at the end of anaerobic glycolysis. Fish muscle LDH activity varies from 100–200 units (μmol substrate converted per g tissue per min) to over 5000 units in tuna white muscle. Muscles of air-breathing fish are at the low end of this spectrum. This trend is also found in other glycolytic enzymes from air-breathers, indicating a general down-regulation of the overall pathway (Table 5.2; see Hochachka et al. 1978b,c, 1979; Hochachka 1979). The demands of the white muscle of swimming fish are fulfilled by anaerobic glycogenolysis, when glycogen is converted to lactate. Comparing the LDH activities in white muscle from water- versus air-breathers, anaerobically powered swimming capacities are suppressed in air-breathing fish (Almeida-Val and Hochachka, in press).

The same pattern occurs in aerobic pathways (Table 5.2). Citrate synthase (CS) is commonly used as an indicator of mitochondrial oxidative metabolism (see Moyes et al. 1992) and it also indicates a down-regulation in the

Table 5.2. Lactate dehydrogenase (LDH), pyruvate kinase (PK), and citrate syntase (CS) activity levels (μmol substrate $min^{-1} g^{-1}$ wet tissue) in selected Amazon fish species. The species *Katsuwonus pelamis* (skipjack tuna) has been included for comparison. (Data compiled from Hochachka and Hulbert 1978; Hochachka et al. 1978b,c, 1979; and Hochachka 1980)

Species tissue	LDH	PK	CS
Lepidosiren paradoxa (lungfish)			
Heart	1777	–	11.5
Synbranchus marmoratus (muçum)			
Heart	871	–	13.5
Arapaima gigas (pirarucu)			
White muscle	260	103	1.71
Red muscle	263	134	3.3
Heart	367	52	10.3
Brain	59	51	0.4
Osteoglossum bicirrhosum (aruanã)			
White muscle	760	536	0.95
Heart	409	144	11.0
Hoplerythrinus unitaeniatus (jejú)			
White muscle	1064	447.5	1.3
Red muscle	810	341	1.4
Hoplias malabaricus (traíra)			
White muscle	576	174	2.0
Red muscle	419	144	3.7
Katsuwonus pelamis (skipjack tuna)			
White muscle	5492	1295	2.1
Red muscle	514	195	21
Heart	449	127	26

muscles of air-breathing fish. White muscle CS levels vary from less than 1 unit in almost all air-breathing fish to approximately 10 units in the white muscles of tuna (Hochachka et al. 1978b; Hochachka 1979; Moyes et al. 1992). The abundance of mitochondria also indicates the same trend (reviewed by Almeida-Val and Hochachka, in press). Thus, performance capacities which are powered by O_2-based metabolism are also inhibited in air-breathing fish.

The heart, on the other hand, is considered a highly oxidative organ in many air-breathers (reviewed by Almeida-Val and Hochachka, in press). The heart of lungfish is characterized by large amounts of mitochondria and oxidative enzymes (Hochachka 1979). The heart of *Arapaima gigas* is as oxidative as lungfish hearts, but it is still less oxidative than hearts of active species such as tuna and trout. Citrate synthase levels indicate similar oxidative capacities in hearts of all air-breathers (Table 5.2). The levels of LDH activity in the heart of *Arapaima* is similar to that present in the hearts of water-breathers. One intriguing fact is that the LDH activity observed in the heart of the lungfish, *Lepidosiren paradoxa*, and in the heart of *Synbranchus marmoratus* is three- to fivefold higher than most fish and mammalian hearts (more than 1000 μmol pyruvate converted per min per gram wet weight; Table 5.2). The LDH activities in the hearts of *Arapaima* and *Osteoglossum* are comparable to that of the heart of tuna, and are half as active as the LDH in lungfish hearts. One explanation for the enormous LDH activities in the Amazon lungfish and in *Synbranchus* relies on their estivation habits during dry periods. At this time, the animal remains completely inactive and thus decreases its whole metabolism. The recovery from estivation is achieved by an efficient back conversion of lactate into pyruvate once O_2 is available to the tissues again. LDH kinetic properties of the heart guarantee this conversion of the accumulated lactate back into pyruvate in both species (Hochachka and Hulbert 1978). Unlike the LDH in its brain (see below), the heart of *Arapaima* LDH exhibits a qualitatively identical picture to *Lepidosiren* and *Synbranchus* heart's LDH, demonstrating that similar solutions to the same problem of lactate accumulation are achieved in the three air-breathers.

Arapaima gigas has levels of brain glycolytic enzymes similar to those in other vertebrates. Interestingly, the LDH in the brain of this fish species is muscle-like, exhibiting almost no inhibition when exposed to high pyruvate concentrations (Hochachka 1980). Actually, brain tissue is almost always susceptible to O_2 depletion and the only good example of anoxia tolerance in the brain occurs in the aquatic turtle, which has a higher glycolytic potential (Lutz et al. 1985). *Arapaima* brain tissue also has a very low oxidative power, having low levels of CS activity (1/20 to 1/40 of other vertebrate brains). It is not known whether *Arapaima* has very low metabolic requirements compared to other vertebrates, and/or if those requirements are simply achieved by anaerobic metabolism (Hochachka 1980). It is most probable that the metabolic requirements of the brains of Amazonian fishes are achieved by both anaerobic and aerobic metabolism. This is supported by the presence of equal concentrations of both LDH loci (A_4 and B_4 isozymes) in brain tissues from almost all of the species that were studied (pers. observ.).

Adaptation to chronic hypoxia in man (generally at high altitudes) is achieved by changes in the metabolic preferences in different tissues optimizing the gain of ATP per mole of O_2 consumed; the preference for glucose reflecting the utilization of glycolysis even in aerobic metabolism (Hochachka 1993). The presence of both LDH M- and H-type in fish brains would guarantee that the accumulated lactate is "washed out" from their brain. The ability to utilize carbohydrate as a preferential fuel for aerobic metabolism, would prevent the high energy costs of turning on and off different metabolic pathways.

5.1.3.1 LDH Isozymes: Channelling Pyruvate

The role of LDH isozymes in fish of the Amazon is worth mentioning because isozyme distribution and its characteristics may indicate metabolic preferences in each tissue (Almeida-Val and Val 1993; Almeida-Val et al. 1993). The main difference among LDH isozymes is that occurring between LDH-B_4 (predominating in aerobic tissues) and LDH-A_4 (predominating in anaerobic tissues). When pyruvate concentrations in aerobic tissues becomes high, LDH-B_4 is inhibited (avoiding the accumulation of lactate in heart for example). In addition, LDH-A_4 activities are not inhibited by high pyruvate concentrations in skeletal muscles, where anaerobic metabolism may cause lactate accumulation without damaging the tissue.

The Amazonian cichlids show a preference for A-type subunits. This behaviour is expected to be common in Amazonian fish due to the chronic hypoxic conditions in which they live. When compared with other teleosts, the LDH electrophoretic patterns for the aerobic tissues in cichlids, such as the heart muscles, show a large predominance of subunit A, with simultaneous reductions in the number of B-like subunits (Fig. 3.9). This tissue distribution appears in other advanced teleosts as well and it is also explained as an adaptation to an environment which has low levels of O_2 (Kettler and Whitt 1986). Relying on anaerobic metabolism is not always the best choice for organisms. However, one fact is clear: some cichlid species are extremely tolerant to anoxic conditions (Soares 1993; pers. observ.) and do not possess either morphological modifications that suggest air-breathing or aquatic surface respiration (ASR) similar to that seen in many Amazon characins (Junk et al. 1983). *Astronotus ocellatus*, for example, can remain over 5 h under anoxic conditions and may retain a high anaerobic power in its tissues.

Despite the low energy yield of anaerobic glycolysis, the activation of this metabolic pathway in response to anoxic conditions is always the first response of "good anaerobes" (hypoxia tolerant), such as aquatic turtles and goldfish, which then depress their metabolism to levels close to zero (Hochachka and Guppy 1987). This also seems to be the case for *Astronotus*, once this species remains almost without activity in the aquaria after 3h of anoxia (Soares 1993; pers. observ.). Aquatic turtles are the best example of this metabolic strategy and also have selected M- or A-type LDH functions in their tissues (Beall and Privitera 1973).

On the other hand, the characin *Colossoma macropomum* is the best example for ASR, using a modified lip extension to skim at the O_2-rich water surface. This species retains a nondivergent pattern of LDH in which both skeletal and heart muscle LDH possesses very similar properties (Almeida-Val et al. 1990, 1991b). Due to the very low Km values for pyruvate and to the inhibition caused by high pyruvate concentrations (10mM), these properties are much more similar to heart-type LDHs than to muscle-type LDHs. Comparing the $Km_{(pyr)}$ values of heart and white muscle LDHs from different teleosts, the LDH in the white muscle of *Colossoma* has Km values similar to those of heart LDHs or the purified LDH-B_4 of other fish species (Table 5.3). The same occurs with its relative *Mylossoma duriventris*. Both species are considered to be tolerant to hypoxia and both are capable of ASR (Soares 1993).

The general trend in species that utilize ASR seems to be the reliance on low and optimized oxidative metabolism. The latter is the only possible alternative for animals adapted to chronic hypoxia because depressing metabolism means cutting down function, which is not possible in many

Table 5.3. LDH $Km_{(pyr)}$ values (mM) of heart and skeletal muscles from selected Amazon fish species. Other teleost's data were included for comparison

Species	$Km_{(pyr)}$		References
	Heart	Muscle	
Hoplias malabaricus[a]	0.33	0.70	French and Hochachka (1978)
Hoplerythrinus unitaeniatus[a]	0.40	1.30	French and Hochachka (1978)
Arapaima gigas[a]	0.15	1.00	French and Hochachka (1978)
Osteoglossum bicirrhosum[a]	0.25	2.30	French and Hochachka (1978)
Mylossoma duriventris[a]	0.041	0.077	Almeida-Val et al. (1991)
Colossoma macropomum[a]	0.056	0.077	Almeida-Val et al. (1991)
Myxine glutinosa	0.45[b]	0.53[b]	Sidell and Beland (1980)
Pseudopleuronectes americanus	0.084[b]	0.56[b]	Markert and Holmes (1969)
Salvelinus fontinalis	0.083	4.80[b]	Wuntch and Goldberg (1970)
Salvelinus namaycush	0.077	3.85[b]	Wuntch and Goldberg (1970)
Melanogramus aeglefinus	0.037[b]	0.33[b]	Sensabaugh and Kaplan (1972)
Salmo salar	0.048[b]	0.40	Gesser and Poupa (1973)
Carassius auratus	0.097[b]	0.355[b]	Wilson (1977); Yamawaki and Tsukuda (1979)

[a] Amazon species.
[b] Tests performed with purified isozymes.

cases (see Hochachka 1993). The ability to make changes in regulatory enzymes is a highly useful adaptation which has occurred, over time, in some Amazonian fishes due to the evolutionary pressure of low O_2 levels in their environment. This seems to be the case for *Colossoma macropomum* which during natural environmental moderate hypoxia, exhibits regulatory changes in its LDH, i.e., the amount of subunits A and B in the heart varies according to the environmental O_2 availability (Fig. 5.1; see Almeida-Val et al. 1990; Almeida-Val and Val 1990). During different periods of the year, when this variation occurs, LDH in the heart has different properties, showing a lower inhibition when exposed to high pyruvate concentrations (Fig. 5.2; see Almeida-Val et al. 1993a). This study was first developed to include different seasons of the year (Almeida-Val 1986). The fact that the electrophoretic pattern of the LDH in the heart showed large amounts of subunit A during part of the year, supported our findings and increased our understanding of the process (Almeida-Val et al. 1990). Additional studies, employing other species belonging to the same family, Serrasalmidae, confirmed this trend further (Almeida-Val et al. 1993).

High sensitivity of LDH to high pyruvate concentration (10 mM) was achieved in skeletal and heart muscles of 12 Amazonian species of Serrasalmidae, piranhas, and pacus (Fig. 5.3; see Almeida-Val et al. 1993).

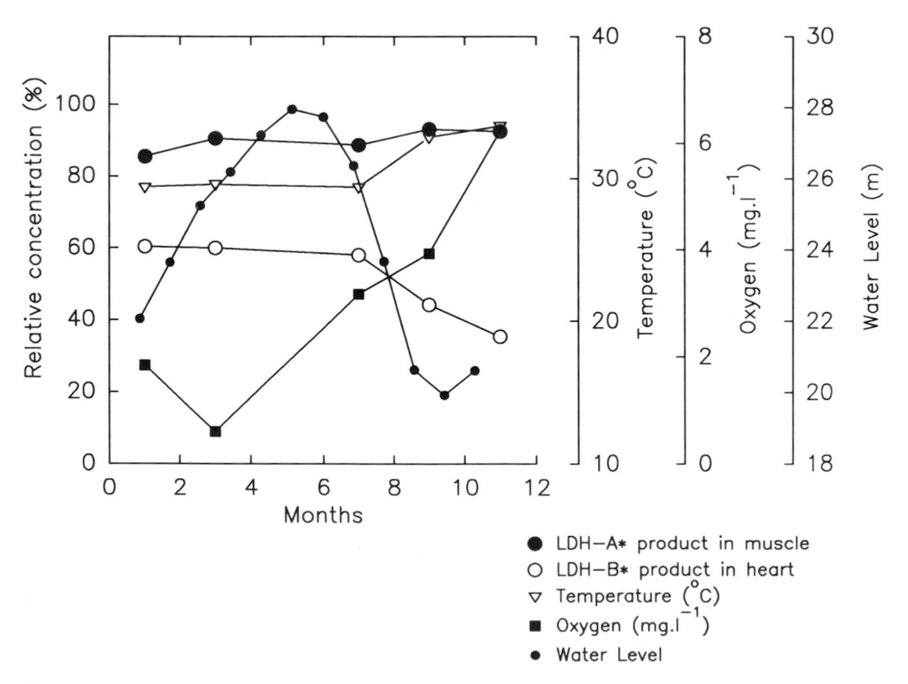

Fig. 5.1. Variation of products of LDH-A* in muscle and LDH-B* in heart of *Colossoma macropomum* during different periods of the year. This variation is related to natural oscillation in dissolved O_2, water level, and temperature

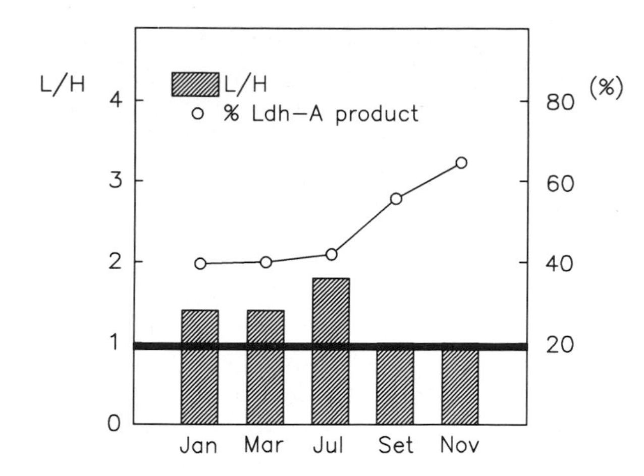

Fig. 5.2. Low/high activity ratios obtained for heart muscle LDH from *Colossoma macropomum* during the year. The increase in relative concentrations of LDH-A* product in the heart of this species is related to a decrease in l/h ratio, i.e., it is related to an activation of anaerobic metabolism during September and November. Low pyruvate concentration = 1 mM; high pyruvate concentration = 10 mM

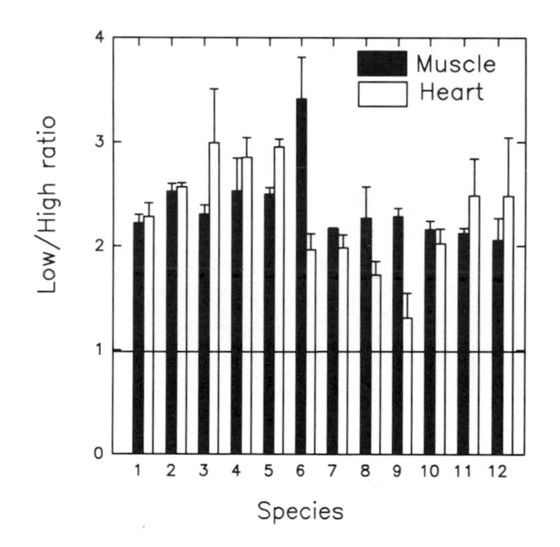

Fig. 5.3. Low/high activity ratios (means and SEM) obtained for skeletal and heart muscle LDH from 12 Serrasalmid species. *1 Myleus rubripinis*; *2 Myleus schomburgkii*; *3 Mylesinus paraschomburgkii*; *4 Metynnis* sp.; *5 Myleus* (*Prosumyleus*) sp.; *6 Serrasalmus* (*Pristobrycon*) *striolatus*; *7 Serrasalmus* (*Pristobrycon*) *eigenmanni*; *8 Serrasalmus* (*Pristobrycon*) sp.; *9 Serrasalmus* (*Serrasalmus*, sp.; *10 Serrasalmus* (*Serrasalmus*) *rhombeus*; *11 Serrasalmus* (*Pygocentrus*) *nattereri*; *12 Catoprion mento*. All species show heart-like functional properties

These species retain heart-type LDH even in skeletal white muscle. The inhibition (low/high ratios for pyruvate) of LDH in these species is much higher than that obtained in other teleost species. Other Amazon teleosts have much lower low/high ratios for the LDH in their muscles than do the serrasalmids which have similar l/h ratio values to those obtained for the LDH in most hearts, especially for *Lepidosiren* and *Synbranchus* hearts (Table 5.4) (reviewed by Almeida-Val et al. 1993).

Studies on the isozyme distribution in five different tissues of these Serrasalmidae species suggested a complex regulatory pattern for LDH isozymes in this family. The predominance of isozyme A_4 was detected in tissues considered predominantly aerobic, such as brain and retina tissue (Table 5.5; see Almeida-Val et al. 1993). As the differential production of LDH-A subunits in *Colossoma macropomum* heart is most likely related to the O_2 availability in the environment (Almeida-Val et al. 1990; Almeida-Val and Val 1990), it is most probable that the differential distribution of loci found in some species of the same group (Serrasalmidae) is due to gene regulation (Almeida-Val et al. 1993). These species, like other Amazonian characins, often combine morphological strategies with biochemical adjustments in order to maintain oxidative processes, adjusting the rates according to their needs and utilizing anaerobic glycolysis only when the activation of this pathway is absolutely necessary.

5.1.3.2 Aerobic Versus Anaerobic Pathways

While enzyme activities give us a good picture of the overall metabolism, their ratios indicate which type of metabolism is preferred as well as which type of regulation is used in different kinds of metabolism (Almeida-Val and Hochachka, in press). Ratios of anaerobic and aerobic capacities seem to be different for air-breathers and water-breathers, confirming the metabolic down-regulation in air-breathers. Useful ratios are: pyruvate kinase/citrate synthase (PK/CS), and lactate dehydrogenase/citrate synthase (LDH/CS). High ratios are present in the brains and hearts of air-breathers (tissues considered to be mainly oxidative), indicating the predominance of anaerobic glycolysis over oxidative pathways (Table 5.6). Compared to water-breathing fish, higher anaerobic potentials are present in the hearts and brains of *Arapaima gigas*, as well as in the heart of both Amazonian and African lungfish.

While heart and brain glycolytic potentials can be viewed as anaerobically powered metabolism and, consequently, as an adaptation to a low O_2 environment; high glycolytic power in muscle represents adaptations for swimming performance (Hochachka and Somero 1984). Thus, PK/CS and LDH/CS can reach values of 2000 in the white muscle of tuna, compared with 153 in the white muscle of *Arapaima* (Table 5.6). An interesting comparison can be made between *Arapaima* and *Hoplerythrinus unitaeniatus*, two Amazon air-breathers (obligate and facultative respectively), under the

Table 5.4. Ratios between LDH activities at low (1 mM) and high (10 mM) pyruvate levels of different tissues from Amazon fish species. (Data compiled from Hochachka et al. 1978a,b,c; Hulbert et al. 1978a,b; and Almeida-Val et al. 1991, 1992)

Species	Tissue[a]	L/H ratio
Hoplias malabaricus	WM	1.13
	RM	1.50
	G	1.91
Hoplerythrinus unitaeniatus	WM	1.07
	RM	1.17
	G	1.39
Osteoglossum bicirrhosum	WM	1.08
	H	1.79
	G	1.38
	K	1.38
	L	0.87
Arapaima gigas	WM	1.43
	RM	1.45
	H	1.50
	G	1.42
	K	1.06
	L	1.30
Lepidosiren paradoxa	H	2.54
Symbranchus marmoratus	H	2.26
Mylossoma duriventris	WM	4.51
	H	3.00
Colossoma macropomum	WM	3.66
	H	3.12
Myleus rubripinis	WM	2.22
	H	2.28
Myleus schomburgkii	WM	2.53
	H	2.57
Mylesinus paraschomburgkii	WM	2.31
	H	2.99
Metynnis sp.	WM	2.53
	H	2.85
Myleus (*Prosumyleus*) sp.	WM	2.50
	H	2.96
Serrasalmus (Pristobrycon) striolatus	WM	3.41
	H	1.97
Serrasalmus (*Pristobrycon*) *eigenmanni*	WM	2.17
	H	1.98
Serrasalmus (*Pristobrycon*) sp.	WM	2.27
	H	1.72
Serrasalmus (*Serrasalmus*) sp.	WM	2.29
	H	1.31
Serrasalmus (*Serrasalmus*) *rhombeus*	WM	2.16
	H	2.02
Serrasalmus (*Pygocentrus*) *nattereri*	WM	2.12
	H	2.49
Catoprion mento	WM	2.06
	H	2.48

[a] WM, white muscle; RM, red muscle; H, heart; L, liver; K, kidney; and G, gill.

Table 5.5. Molar ratio[a] of LDH B/A (−) duplicate gene expression in 12 species of Amazon serrasalmids (piranhas and pacus).[b] (Data from Almeida-Val et al. 1992)

Species	SM	H	L	R	B
C. macropomum					
Mean	0.21	1.78	2.48	1.23	0.85
Log	(−0.67)	(0.25)	(0.40)	(0.09)	(−0.07)
Min-max	0.18–0.21	1.41–2.38	2.00–2.83	1.00–1.41	0.71–1.00
M. duriventris					
Mean	0.39	1.16	2.20	0.93	0.79
Log	(−0.41)	(0.07)	(0.34)	(−0.03)	(−0.10)
Min-max	0.21–0.60	1.00–1.41	1.41–2.83	0.84–1.00	0.60–1.00
P. brachypomum					
Mean	0.45	2.52	1.81	0.73	1.00
Log	(−0.34)	(0.40)	(0.26)	(−0.14)	(0.00)
Min-max	0.41–0.60	1.68–4.00	1.19–2.38	0.60–0.84	1.00–1.00
M. rubripinis					
Mean	0.69	2.19	1.30	1.68	1.21
Log	(−0.16)	(0.34)	(0.11)	(0.23)	(0.08)
Min-max	0.60–0.84	2.00–2.38	1.19–1.41	1.68–1.68	1.00–1.41
M. paraschomburgkii					
Mean	0.72	1.95	1.97	0.63	1.23
Log	(−0.14)	(0.29)	(0.29)	(−0.20)	(0.09)
Min-max	0.60–1.00	1.68–2.38	1.41–2.38	0.60–0.71	1.19–1.41
Metynnis sp.					
Mean	0.70	1.82	2.00	0.82	1.00
Log	(−0.15)	(0.26)	(0.30)	(−0.09)	(0.00)
Min-max	0.60–0.84	1.41–2.00	2.00–2.00	0.60–1.00	1.00–1.00
S. (*Pristobrycon*) *striolatus*					
Mean	0.49	1.20	1.62	1.51	1.10
Log	(−0.31)	(0.08)	(0.21)	(0.18)	(0.04)
Min-max	0.42–0.71	1.00–1.41	1.00–2.00	1.00–2.00	1.00–1.41
S. (*Pristobrycon*) *eigenmanni*					
Mean	0.69	1.78	2.18	0.84	0.91
Log	(−0.16)	(0.25)	(0.34)	(−0.08)	(−0.04)
Min-max	0.50–1.00	1.41–2.38	1.68–2.83	0.84–0.84	0.91–1.00
S. (*Pristobrycon*) sp.					
Mean	0.50	1.44	0.80	0.89	0.91
Log	(−0.30)	(0.16)	(−0.10)	(−0.05)	(−0.04)
Min-max	0.42–0.60	1.00–2.00	0.71–0.84	0.71–1.00	0.84–1.00
S. (*Serrasalmus*) sp.					
Mean	0.59	1.04	1.17	1.08	0.91
Log	(−0.23)	(0.02)	(0.07)	(0.03)	(−0.04)
Min-max	0.50–0.71	1.00–1.19	1.00–1.68	1.00–1.41	0.84–1.00
S. (*Serrasalmus*) *rhombeus*					
Mean	0.50	0.96	2.11	1.28	1.08
Log	(−0.30)	(−0.02)	(0.33)	(0.11)	(0.03)
Min-max	0.42–0.60	0.84–1.00	1.68–2.83	1.00–2.00	1.00–1.41
S. (*Pygocentrus*) *nattereri*					
Mean	0.50	2.41	1.71	0.94	1.00
Log	(−0.31)	(0.38)	(0.23)	(−0.03)	(0.00)
Min-max	0.42–0.60	2.00–2.38	1.41–2.38	0.84–1.00	1.00–1.00

[a] Molar ratios are presented as means (log) of 6 different measurements in each tissue.
[b] SM, skeletal muscle; H, heart; L, liver; R, retina; B, brain.

Table 5.6. Comparative ratios of enzymes from anaerobic and aerobic pathways in selected Amazon species. Air- and water-breathing fishes are compared. The species *Protopterus aethiopicus* (African lungfish) and *Katsuwonus pelamis* (skipjack tuna) have been included for comparison. (Data compiled from Hochachka and Hulbert 1978; Hochachka et al. 1978b,c, 1979; Hochachka 1980; Dunn et al. 1983)

Species	Tissue	LDH/CS	PK/CS
Arapaima gigas			
	Heart	37	5
	White muscle	153	60
	Red muscle	80	41
	Brain	140	128
Lepidosiren paradoxa			
	Heart	102	–
Protopterus aethiopicus			
	Heart	33	6
	White muscle	306	118
	Brain	58	40
Osteoglossum bicirrhosum			
	Heart	37	13
	White muscle	800	593
Hoplerythrinus unitaeniatus			
	White muscle	819	343
	Red muscle	579	244
Hoplias malabaricus			
	White muscle	288	87
	Red muscle	113	39
Synbranchus marmoratus			
	Heart	65	–
Katsuwonus pelamis			
	Heart	17	5
	White muscle	2034	480
	Red muscle	25	10

same environmental pressure. Low anaerobic capacities are present in the white muscle of *Arapaima* compared to a high LDH/CS rate (more than 800) in *Hoplerythrinus*. This fact is explained on the basis of their swimming performance and life-styles (Almeida-Val and Hochachka, in press).

Again, these enzymatic rates substantiate both the high oxidative characteristics found in the heart of *Arapaima* (PK/CS value is similar to the values obtained for tuna hearts) and the high glycolytic potential (LDH/CS = 148 and PK/CS = 128; see Table 5.6) of its brain. These adjustments were probably made during the evolution of the enzymes, together with changes in structural and regulatory genes to provide metabolic regulation in these species. In order to estimate the immediate effects of exposure to hypoxia, however, some metabolic adjustments may need to be evaluated.

5.2 Effects of Hypoxia

Hypoxia-induced responses in aquatic animals are well-defined and may be categorized as: (1) oxygen regulators – in the face of hypoxia, these fish do not change the amount of O_2 that they take in but rather maintain a constant, low rate of O_2 consumption. (2) Oxygen conformers – these fish gradually reduce their O_2 consumption according to the decrease in the amount of O_2 available in the water.

Among air-breathing fish, two respiratory patterns can be recognized under hypoxia; (1) some species have well-developed branchial arches and exhibit air-breathing only when exposed to very low levels of dissolved O_2; and (2) other species have functional lungs and breathe air at a higher rate than they breathe water, even under normoxic conditions (Glass 1992, see Chap. 4). The latter category includes the obligatory air-breathers whose responses to hypoxic situations have been evaluated at two different levels: (a) exposure to hypoxia, and (b) breath-holding during diving.

The Amazon *Hoplerythrinus unitaeniatus* is included in the first category and increases air-breathing rate according to O_2 depletion in the water (Fig. 5.4; see Stevens and Holeton 1978b). On the other hand, *Arapaima* is an obligatory air-breather and does not change the typical 80% air-breathing rate even in normoxic waters; but it does decrease O_2 uptake when submitted to low O_2 levels in both water or air (Fig. 5.5 and 5.6; see Stevens and Holeton 1978a).

The rudimentary gills of the Amazonian lungfish (*Lepidosiren paradoxa*) are similar to the gills of the African lungfish, *Protopterus* spp. None of the

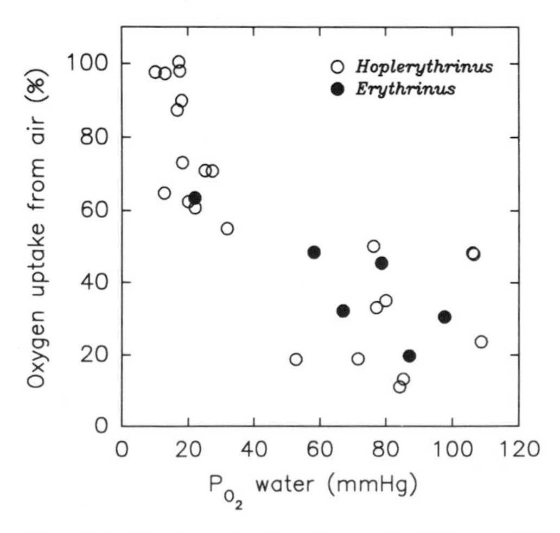

Fig. 5.4. Oxygen uptake from air (%) at different concentrations of O_2 in the water. There is an increase in air-breathing as P_{O_2} decreases in the water both for *Hoplerythrinus* and *Erythrinus*. (Redrawn from Stevens and Holleton 1978b)

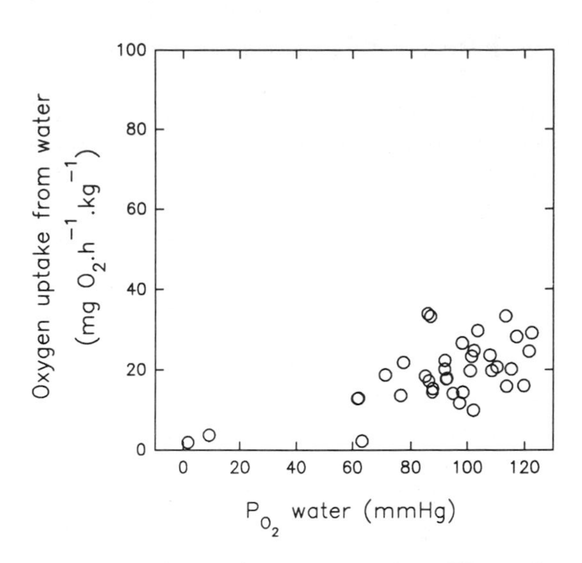

Fig. 5.5. Fraction of O_2 uptaken from water by pirarucu (*Arapaima gigas*) at different O_2 pressures in the water. As water O_2 decreases, water O_2 uptake also decreases. (Redrawn from Stevens and Holleton 1978a)

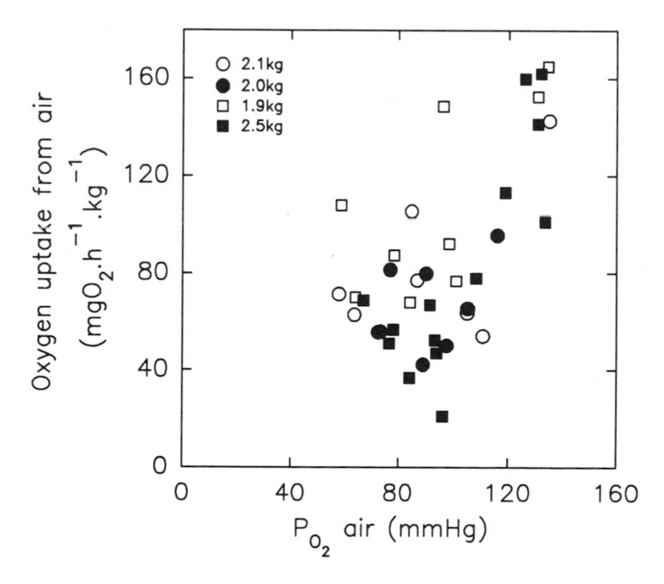

Fig. 5.6. Oxygen uptake from air from 4 specimens (different sizes) of pirarucu, as affected by changes in air P_{O_2}. There is a decrease in metabolic rate according to the decrease in air P_{O_2}. (After Stevens and Holeton 1978a)

specimens of both genera can survive if denied access to the water's surface, unlike the Australian lungfish (*Neoceratodus forsteri*) which is a facultative-breather (Johansen and Lenfant 1968; McMahon 1970). Even though hypoxic exposure or submergence studies have not been done with *Lepidosiren*, some characteristics of *Protopterus* could help us to understand what would happen with *Lepidosiren* when exposed to hypoxia. As mentioned earlier, exposure to hypoxia induces an increase in ventilation (increasing O_2 uptake) in *Protopterus* both through the lungs and the gills (Fishman et al. 1989). Reviewing the physiology of diving in lungfish, Fishman et al. (1989) pointed out that the lungfish does not respond to hypoxia in water and, consequently, the dipnoan lungfish does not increase its breathing frequency as holostean air-breathers do. Instead, it has been suggested that they depress gill ventilation during aquatic hypoxia (Johansen et al. 1970).

The metabolic changes which occur during exposure to hypoxia has been summarized by Hochachka (1988b). In his studies, the author examined the functions and structure of the epaxial muscle in lungfish. When exposed to hypoxia of 5 torr or less, this tissue relies on glycolysis to generate 50% or more of its total, although low, ATP turnover rates. The energy levels in lungfish muscles do not vary and there is no glycolytic activation to compensate for drops in the concentration of O_2 (Dunn et al. 1983). According to Hochachka (1988b), this tissue goes into metabolic arrest in order to survive periods of hypoxia.

Amazonian water-breathers exposed to severe hypoxia for 48 h demonstrate an initial increase in O_2 uptake and, after the first 3 h of exposure to hypoxia, the metabolic rate decreases to about 50% of their initial O_2 intake (Fig. 5.7; Almeida-Val and Hochachka, unpubl. data).

5.2.1 Metabolites

Metabolites change qualitatively and quantitatively in different tissues when the whole animal is subjected to hypoxia: lactate accumulates in the tissues, glucose is mobilized from liver glycogen in order to supply anaerobic glycolysis in several tissues, and energy turnover decreases initially and later returns to normal (Hochachka 1988b). In "good anaerobes", such as aquatic turtles, optimum energy turnover is maintained in the central nervous systems (CNS), i.e., the animal is able to maintain ATP concentrations throughout anoxia (Lutz et al. 1984, 1985).

The water-breather *Hoplias malabaricus* is much more tolerant to anoxic conditions than the air-breather *Hoplerythrinus unitaeniatus* (Driedzic et al. 1978). Both species exhibit similar energy levels in their hearts but differ greatly in the amount of glycogen sustained by their hearts. Exposure to acute anoxia can be sustained by *Hoplias* for 6 h. During this time, lactate accumulates and glycogen is depleted, indicating the complete reliance on

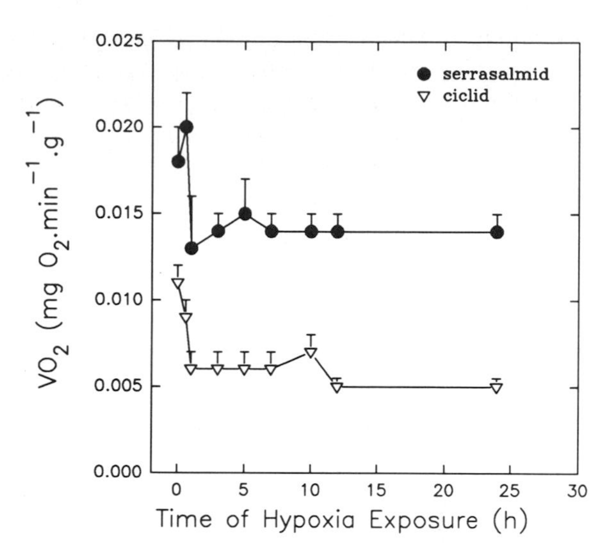

Fig. 5.7. Water O_2 uptake of *Serrasalmus* sp. and *Cichlasoma* sp. 24 h after exposure to severe hypoxia (30 mm Hg). The decrease in metabolic rate is amplified in cichlid compared with serrasalmid. (Data from Almeida-Val and Hochachka, unpubl.)

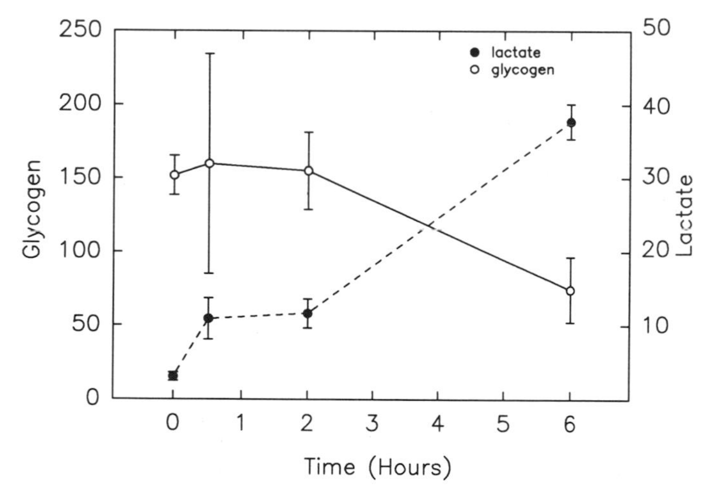

Fig. 5.8. Levels of glycogen (μmol glucose/g wet tissue) and lactate (μmol/g wet tissue) in the heart of *Hoplias* exposed to acute anoxia. (Redrawn from Driedzic et al. 1978)

anaerobic glycolysis (Fig. 5.8; see Driedzic et al. 1978). The heart of *Hoplias* is also tolerant to cyanide perfusion, being able to sustain its function for up to 23 min. *Hoplerythrinus* heart, on the other hand, can survive for only 3 min when perfused with the same drug. As *Hoplias* is unable to breathe air, its heart must be adapted to rely on anaerobic metabolism during O_2

depletion. We are unable to assume that *Hoplias* is as a "good anaerobe" as the aquatic turtle because at the end of the 6 h of anoxia, the cardiac output of *Hoplias* is extremely reduced (Driedzic et al. 1978). Also, no adenylate measurements have been made on *Hoplias* after exposure to anoxic conditions to verify changes in the energy turnover.

Several authors have determined that *Hoplias malabaricus* is tolerant to hypoxia (Driedzic et al. 1978; Hochachka et al. 1978b) based on the fact that this species accumulates large amounts of glycogen in its tissues (red muscle, liver, and heart). Based on physiological measurements, Rantin and Johansen (1984) also suggested that this species is tolerant to hypoxia. However, the reliance on anaerobic metabolism does not conserve the high energy necessary for vigorous swimming, which is part of this predator life. This species can either resort to ASR when O_2 in the water is depleted or remain quietly at the bottom of its aquarium as if the metabolism had been fully depressed (Soares 1993). Since *Hoplias malabaricus* uses O_2 more efficiently, glucose should be used as aerobic fuel by species which are tolerant to hypoxia (as does the hypoxia-adapted man) (Hochachka 1993). The presence of large amounts of glycogen in the aerobic organs of *Hoplias* is a good indication that this fish species preferentially uses carbohydrate rather than lipid as fuel for aerobic metabolism.

We have suggested that choosing metabolic suppression and maintaining oxidative capacities despite chronic O_2 limitation, is probably a coadaptation along with the acquisition of alternate morphological and behavioural defense mechanisms in Amazon fishes (Almeida-Val et al. 1993). Our studies of LDH regulation in *Colossoma macropomum* during different environmental conditions showed that lip extension and ASR are not the only mechanisms utilized by this species. Experiments exposing the whole animal to hypoxia showed that the development of lips occurs only when the fish has been submitted to extremely severe hypoxia ($0.5\,mg\,O_2 l^{-1}$) for over 2 h. Before lip development, lactate accumulates in the blood. After lip extension, blood lactate contents do not decrease immediately but tend to stabilize (Fig. 5.9). During the recovery period, after returning the fish to normoxia, a decrease in lactate production, with lip retraction, is observed within the next 2 h (Fig. 5.9). *Colossoma* LDH is inhibited in vitro by low pHs and high pyruvate concentrations (Almeida-Val et al. 1991b). At the pyruvate branching point, these properties help to prevent lactate accumulation and drive pyruvate through the Krebs cycle, once O_2 is available again.

5.2.2 Tissue Metabolic Regulation

Surviving hypoxia can be achieved only by the regulation of metabolism within each separate tissue. Because of this, several adaptive features create different metabolic arrangements in different organs. For example, the liver must store large amounts of glycogen in order to supply fuel for emergencies,

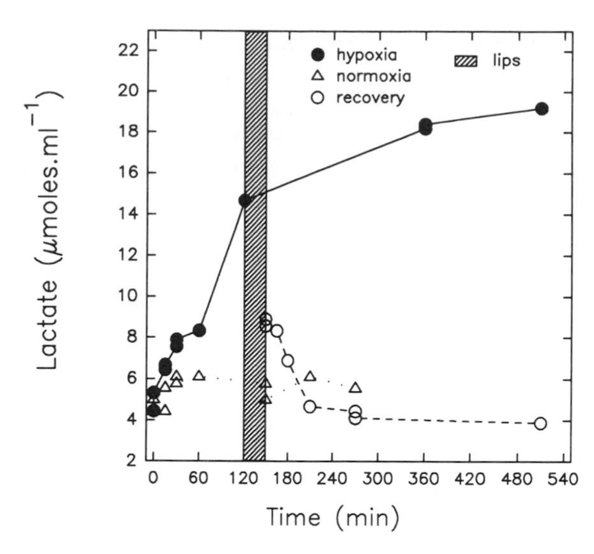

Fig. 5.9. Lactate levels in plasma (μmol/ml) from *Colossoma macropomum* under hypoxic ($O_2 < 0.5$ mg/l) and normoxic (supersaturation) conditions. (Almeida-Val et al. 1993)

and the heart must be able to utilize the lipids or even the glycogen which have accumulated in its tissue without oscillating energy levels and heart outputs.

As previously mentioned, different organs of *Arapaima* sustain different enzymatic compositions (isozyme predominance, differential enzyme activities). In this species, the citrate synthase activity in the heart is approximately 25 times higher than in brain, and 10 times higher than in kidney, liver, or white muscle. Because the heart has the highest ATP turnover rate pcrgram of tissue, it is probably the main O_2 consumer (Hochachka 1980). Kidney tissue is the only other tissue in *Arapaima* which could, consume almost as much O_2 as the heart because of its metabolic characteristics (Hochachka et al. 1978a; Storey et al. 1978). The enormous size of the *Arapaima* kidney (Fig. 4.5), about 3.5-fold larger than its water-breathing relative *Osteoglossum*, and the presence of larger amounts of mitochondria, contribute to a higher energy consumption during O_2-depleted situations (diving, for example). In this particular case, the increased formation of ammonia and glucose contribute to the ionic balance (reducing pH perturbations) and glucose homeostasis of the whole organism. This is a good example of tissue regulation under metabolic depression and a similar situation may also exist in lungfish (reviewed by Almeida-Val and Hochachka, in press).

5.3 Temperature Effects

Temperature affects a variety of natural processes. For example, in some organisms, their metabolic rates may be affected by temperature changes as well as by their own geographical distribution (Prosser 1991). Metabolic compensation (Precht et al. 1973) or acclimatization (Hazel and Prosser 1974) can be achieved in the organism by changes at the metabolic level. Endotherms usually keep their body temperature constant and independent from the environmental temperature oscillations. Ectotherms, on the other hand, usually adjust their body temperatures according to the ambient temperature (Hochachka and Somero 1973).

Almost all fish are ectotherms. There are at least five different processes by which these fish can compensate for temperature changes: (1) changes in substrate concentrations and/or products in a single pathway; (2) changes in modulator concentrations affecting the enzymatic reactions; (3) changes in the enzyme conformation that affects its substrate affinity (Km) and its velocity (Vmax); (4) quantitative changes in the enzyme synthesis by gene regulation; and (5) qualitative changes in isozymes. All these mechanisms strongly influence the thermal acclimatization or acclimation of organisms in general (Hochachka and Somero 1984).

Because long-term temperature changes are not observed in the Amazon basin (see Chap. 2), its effectiveness in inducing metabolic change is reduced. Temperature drops to 14 °C are uncommon during the year and occur in only 2 or 3 days in June. Short-term oscillations (hours) are mainly compensated for by behavioural responses in ectotherms, which avoid daily oscillations by migrating to different microhabitats (Hochachka and Somero 1984). Long-term climatic changes, however, provide enough time for phenotypic adjustments via gene regulation, or for changes in the quantities or the types of enzymes. In addition, genetic modifications may occur over the course of many generations through geological time. These changes would include structural changes in enzymes and/or isozymes and, consequently, changes in their physical and chemical catalytic characteristics. The temperatures of the Amazon region may have changed drastically during the geological formation of the basin, or during glacial ages, when many species were confined to smaller, milder habitats. These habitats are known as refuges (Gentilli 1949; Vanzolini 1973). Expansions of the aquatic environment during interglacial ages allowed for the colonization of new habitats. This process may have forced the animals to acclimatize, adjusting their metabolisms to the new situation. Some enzyme properties, such as thermostability, are intrinsic properties of the molecules which are rarely influenced by rapid environmental changes.

5.3.1 Isozyme Thermostability

Most isozymes can be characterized by their differences in thermostability. For example, the isozyme encoded at the sMDH-B* locus is much more thermolabile than the isozyme encoded at sMDH-A* locus. In the case of LDH, LDH-A_4 isozymes are much more thermostable than LDH-B_4 isozymes. These differences help in establishing homology between different loci. Opposite results for LDH were found in two Amazonian Siluriformes; *Hoplosternum littorale* and *Brachyplatystoma filamentosum*. The enzymes (predominantly LDH-A_4) in their skeletal muscle are more thermostable than the enzymes in the heart muscle (predominantey LDH-B_4) (D'Ávila-Limeira 1989). Similar results have been described for one Cypriniform (Hauss 1975) and one amphibian species (Goldberg and Wuntch 1967).

While searching for homology among orthologous LDH loci in the fishes of the Amazon (D'Ávila-Limeira 1989), we found interesting differences among the inactivation temperatures of LDH-B* loci in different species. Among the species analyzed, *Potamotrygon motoro*, Rajiformes has the smallest inactivation temperature (55 °C) while *Plagioscion squamisissimus*, a perciform, has one of the highest inactivation temperatures (70 °C) (Fig. 5.10). It is interesting to observed that some families among the Characiforms have an inactivation temperature averaging 82 °C.

The above-mentioned heat stability for LDH is observed in species of Curimatidae (Fig. 5.10). This family, as already stated (Chap. 3), has a peculiar history. They colonized the Amazon basin before the Andes uplift. Thus, the Curimatidae may have endured temperature oscillations during their evolution; time enough to allow for structural changes in proteins. These evolutionary pressures may have influenced the types of catalytic responses of the LDH to different temperatures, reflecting some adaptation to environments where the temperature is high.

The main role of enzymes is to decrease the energy of activation of a chemical reaction, enabling it to occur at moderate temperatures. According to Hochachka and Somero (1984), the flexibility of the enzyme structure probably occurs during binding events; this modification is known as the "induced fit" model (Koshland 1973) or "hand and glove" model (Hochachka and Somero 1984). Change in enzyme conformation will be accompanied by changes in the energy inputs or outputs which are associated with conformational changes and probably responsible for changes in the energy of activation.

Wilson et al. (1964), who studied the thermostability of LDH-B_4 isozymes in 55 different species of different vertebrates (from lower vertebrates to mammals), suggested that the increases in the heat stability of this isozyme is related to the phylogenetic status of the species. Thus, in their diagram, mammals can be found in one branch of the phylogenetic tree and reptiles and birds in the other branch; the latter having LDHs which are 20 °C more

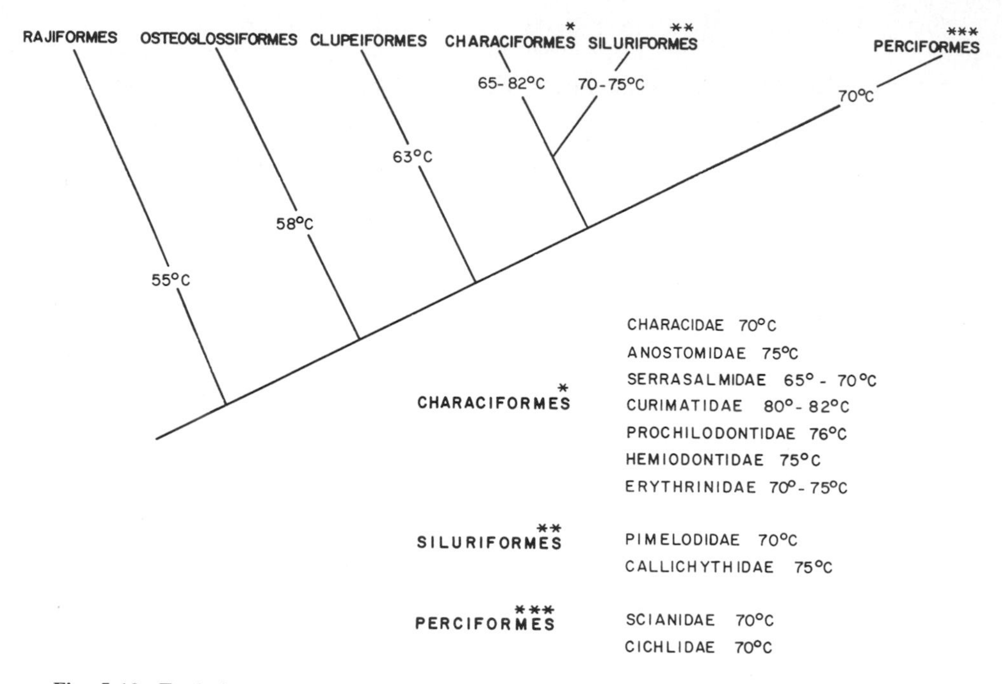

Fig. 5.10. Evolutionary relationship among main fish orders (families in detail) showing the lowest inactivation temperature of LDH isozymes. Less specialized orders present lower thermal resistance. (Data from D'Avila-Limeira 1989; Almeida-Val, unpubl.)

resistant than those of fish. In this case, it is possible that the LDH thermal properties are also related to body temperature.

Adaptations to higher temperatures may induce increases in the structural stability of proteins (Hochachka and Somero 1984). One of the consequences of increasing structural stability by increasing the number of bonds (weak bonds) is the decrease in the catalytic efficiency (increasing the energy of activation). Furthermore, there is a strong correlation between body temperature and protein thermal stability. Thus, there is a loss in catalytic efficiency as protein stability is increased. If the body temperature is related to heat stability, experiments with fish who are exposed to different thermal regimes should give some answers regarding this relationship.

Some Amazonian fish species exposed to different temperatures in the laboratory had the LDH heat stabilities analyzed in both their heart and muscle tissues (Farias 1992). Thermal stability indexes were determined after measurements of enzyme activity at different periods of heat exposure. This index was a measure of thermal resistance and is correlated with body temperature (acclimation temperature). Heart LDH from *C. macropomum* increases in stability as the body temperature increases. Similar results have been obtained for the LDH in *Hoplosternun littorale* hearts (Fig. 5.11). The stabilities of the LDH in the skeletal muscles of these animals are, on

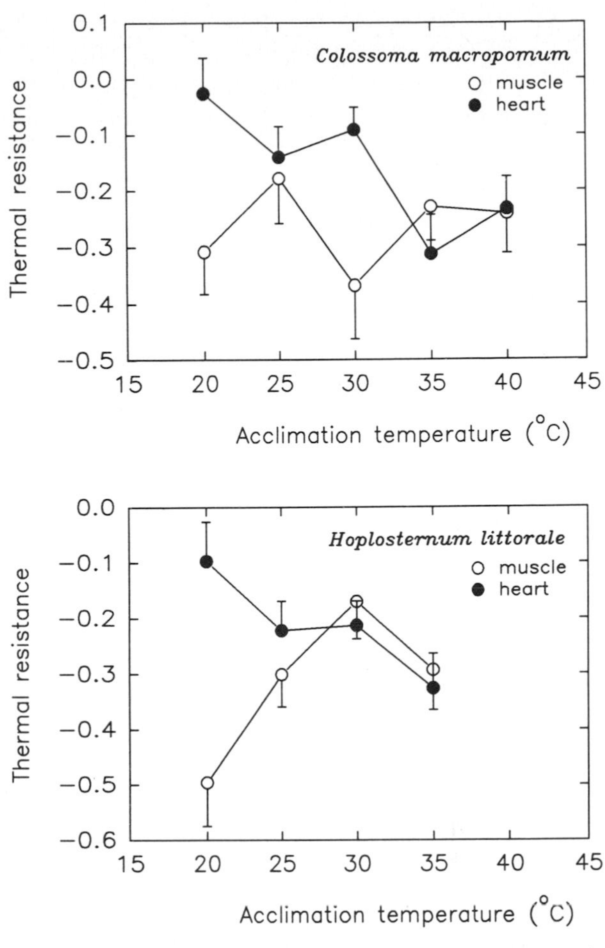

Fig. 5.11. Thermal resistance as affected by temperature of acclimation in heart and muscle LDH from *Colossoma macropomum* and *Hoplosternum littorale*. (Data from Farias and Almeida-Val, unpubl.)

the other hand, inversely correlated with body temperature, i.e., when the acclimation temperature increases thermal resistance decreases. The acclimation temperature effect, however, is stressed in *H. littorale* when compared to the LDH in the muscles of *C. macropomum*, which appears to be more stable (Fig. 5.11). During the acclimation of *Hoplosternum littorale*, the same reverse pattern is repeated at 30 and 35 °C in spite of the similarity of the inactivation values of both heart and muscle LDHs. It is clear that the acclimatization process affects thermal resistance and that, at higher temperatures, the homotetramers B_4 (heart isozymes) loose their characteristic thermostability and tend to behave like the homotetramer A_4 under heat exposure.

Nevertheless, after 1 h of heat exposure in vitro, the enzymes of both tissues in both species do not lose more than 70% of their initial activity (Farias 1992). Actually, selection for higher thermostability patterns may be a rule in the evolution of LDH in Amazonian fish. The differences in their respiratory patterns and strategies to deal with metabolic O_2 deficiencies must have imposed the selection of different molecular properties during the evolution of the enzymes/isozymes in these species, thus providing different heat stabilities.

5.3.2 Km Versus Temperature

For the fish of the Amazon, life in a higher temperature range may have imposed the necessity for more stable proteins. It is, however, important to know whether the catalytic efficiency was affected during the evolution of those species. Thermal modulation by enzymes can be divided into two categories: positive and negative. Positive thermal modulation occurs when a direct relationship is observed between temperature and Km, i.e., the higher the temperature, the higher the Km. Negative thermal modulation induces a decrease in Km according to an increase in temperature. Some enzymes or isozymes exhibit both kinds of thermal modulation, showing a V-shaped curve according to temperature increases. At the bottom of the curve is the minimum Km value, i.e., the maximum enzyme-substrate affinity value. This point generally occurs near the acclimatization temperature and may vary depending on whether the same species is cold- or warm-acclimated (Hochachka and Somero 1968, 1973, 1984).

In the case of Amazonian fish, the minimum Km for enzymes should lie within the range of 25–30 °C. In fact, the first measurements by Hochachka and Somero (1968) show that the LDH in lungfish muscle has its minimum Km in the range of 30–35 °C (Fig. 5.12). Comparing the LDH of lungfish muscle and that of tuna muscle, one can suggest that the acclimatization temperature is very important for metabolic functions. It is important to keep the Km values close to the substrate concentrations in the tissues, otherwise the enzyme will not be able to work, or it will require a much higher activation energy to reach the optimum rate.

5.3.3 Temperature and pH: Combined Effects

Intracellular pHs vary according to temperature changes (Reeves 1977; Somero 1981). Thus, the combination of pH and temperature has a significant effect on the regulation of enzyme activity in the organism. The temperature modulation of LDH in trout liver, for example, is stressed in alkaline pHs (Hochacka and Lewis 1971). Because it is hazardous for organisms to vary their intracellular pHs, it is important to have metabolic

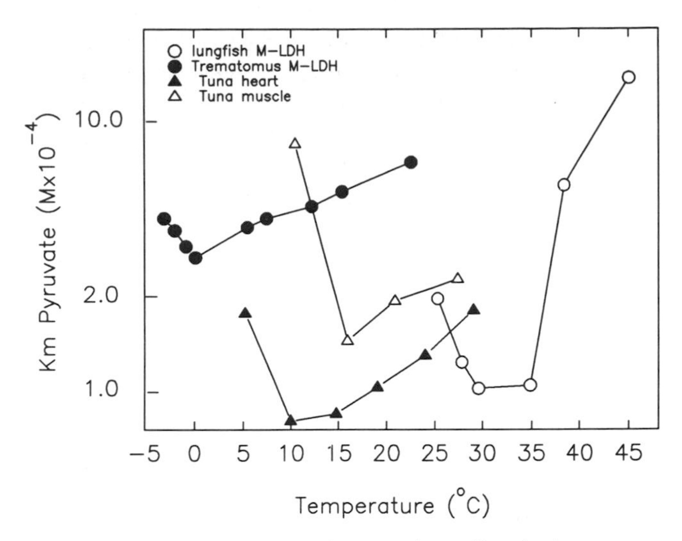

Fig. 5.12. LDH $Km_{(pyr)}$ values as affected by temperature from species acclimatized at extreme thermal regimes. Minimum Km values for each species occur around environmental temperature. (Redrawn from Hochachka and Somero 1968)

regulation which will decrease the effect of temperature on physiological pHs and therefore on the Km values of the enzymes.

The LDH of Amazonian fish is similarily affected by temperature and pH changes despite the small oscillation of temperature in the Amazon basin. The temperature effects on LDH from muscle and heart tissue of Amazon species (Almeida-Val et al. 1991b) are also minimized when pHs are close to the physiological norm (Fig 5.13). This effect is stressed in the LDH of *Colossoma macropomum* hearts. This tissue exhibits no variation in substrate saturation in response to temperature when the pH is 7.5 as compared to *Mylossoma duriventris* (Fig. 5.14). In addition, high temperatures, such as those found in the Amazon, should increase the energy requirements of the animal and, consequently, should increase enzyme Km values (decreasing enzyme-substrate affinities).

Compared to *Colossoma macropomum* LDH, this enzyme in the species *Mylossoma duriventris* is much more influenced by temperature (Fig. 5.15), especially in the heart muscle. However, the effects that pH has on LDH is similar for both species. Thus, the effects of temperature on the LDH in *M. duriventris*, at pH 7.5 or 6.0, are much lower than those when the pH is 8.5. Interestingly, the LDH in the heart muscle of *C. macropomum* is not affected by temperature at all (Table 5.7). Keeping Km values constant at physiological pHs in the LDH of heart is almost mandatory because heart muscle is not able to sustain a high rate of anaerobic glycolysis. High affinity values (i.e., low Km values) for LDH assures a low anaerobic glycolysis rate in the heart of both species. On the other hand, sustaining high anaerobic

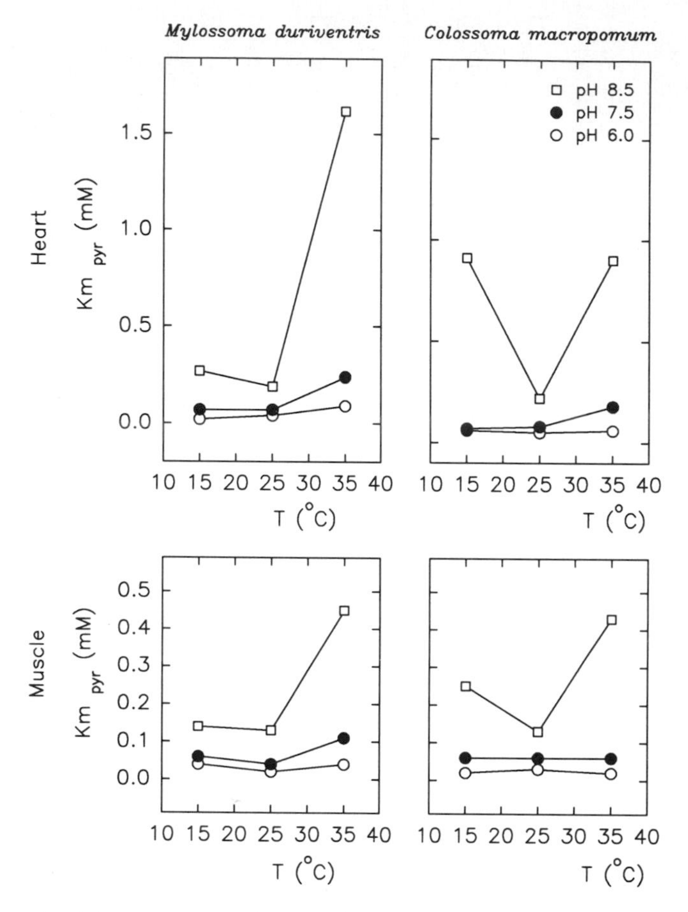

Fig. 5.13. The effect of pH and temperature on LDH $Km_{(pyr)}$ in muscle and heart of two selected Amazon species (*Colossoma macropomum* and *Mylossoma duriventris*). The effect of temperature on $Km_{(pyr)}$ is minimal at physiological pH (6.0 and 7.5)

glycolysis, namely during intense exercise when O_2 is depleted, is normal in skeletal muscle. It is surprising that Km values for the LDH in the skeletal muscle of both species are very low (heart-like) compared to other teleost species. Again, the maintenance of the metabolism in a low oxidative range would explain this fact. This seems to be the case for the whole group to which *Colossoma* belongs (Serrasalmidae family), as stated above.

Another temperature and pH-dependent parameter is the inhibtion of the enzyme by increasing substrate concentrations. The effects of temperature and pH over pyruvate inhibition have been determined for both *C. macropomum* and *M. duriventris* (Fig. 5.16). As we can see, at a lower pH the inhibition of LDH is much higher than at a pH ranging between 7.5 and 8.5 in both the heart and skeletal muscles of both species. Actually, both species

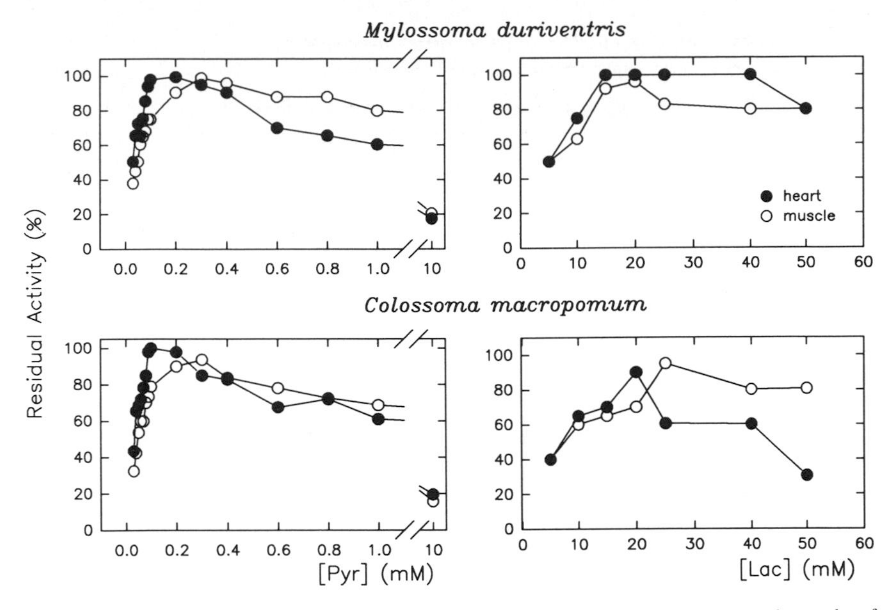

Fig. 5.14. Comparative LDH saturating curves (pyruvate reduction) from heart and muscle of two selected Amazon species (*Mylossoma duriventris* and *Colossoma macropomum*)

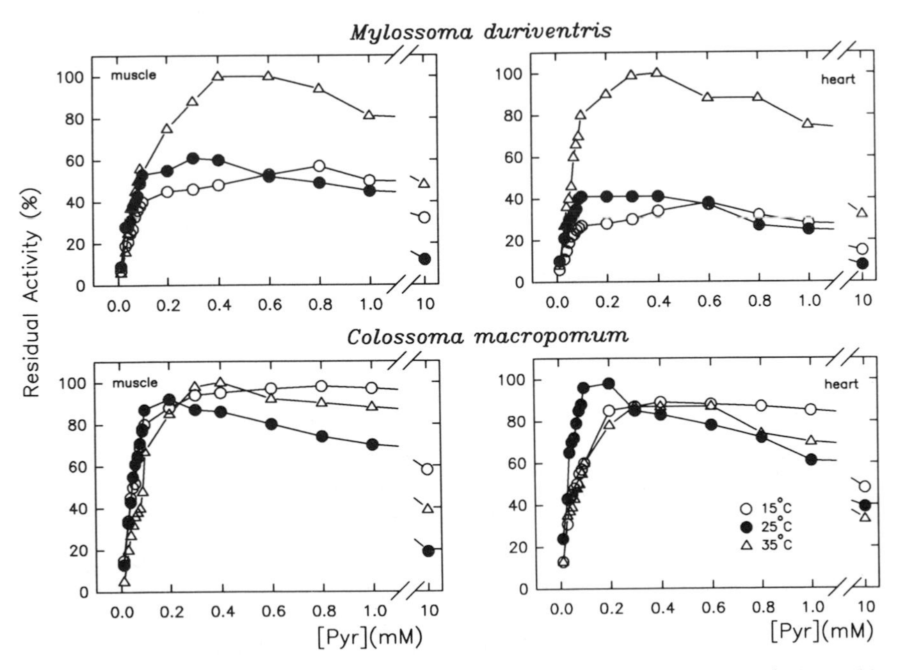

Fig. 5.15. Temperature effect on LDH saturating curves (pyruvate reduction) from *M. duriventris* and *C. macropomum*. Note that heart and muscle LDH from *C. macropomum* is temperature-independent

Table 5.7. LDH $Km_{(pyr)}$ values at different pHs and temperatures for skeletal and heart muscles from two Amazon serrasalmids. (Data from Almeida-Val 1986)

Skeletal muscle LDH Km_{pyr} values (mM)

pH	*Mylossoma duriventris*			*Colossoma macropomum*		
	Temperatures (°C)			Temperatures (°C)		
	15	25	35	15	25	35
6.0	0.02	0.04	0.09	0.06	0.05	0.06
7.5	0.07	0.07	0.24	0.07	0.06	0.18
8.5	0.27	0.19	1.62	0.91	0.22	0.90

Heart muscle LDH Km_{pyr} values (mM)

pH	*Mylossoma duriventris*			*Colossoma macropomum*		
	15	25	35	15	25	35
6.0	0.04	0.02	0.04	0.02	0.03	0.02
7.5	0.06	0.04	0.11	0.06	0.06	0.06
8.5	0.14	0.13	0.45	0.25	0.13	0.43

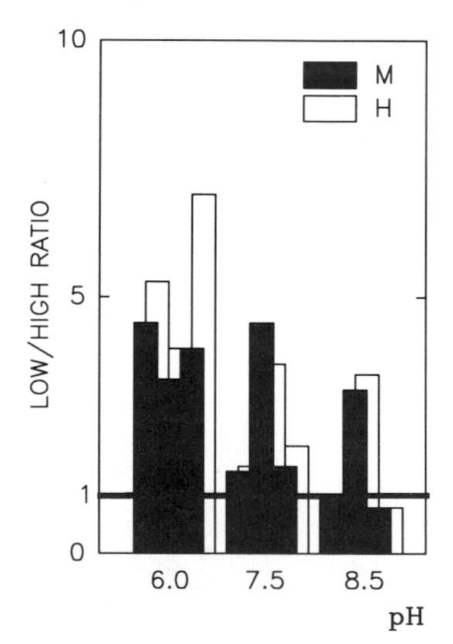

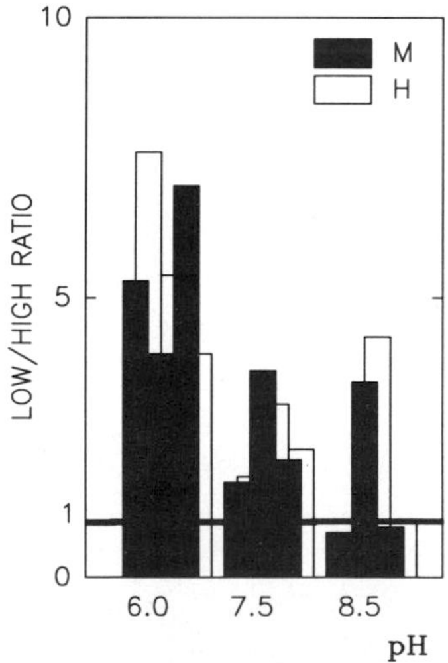

Fig. 5.16. L/H activity ratios of LDH from heart and muscle of two *M. duriventris* and *C. macropomum* as affected by pH and temperature (each *bar set* corresponds to the sequence: 15, 25, and 35 °C)

exhibit very similar responses regarding pyruvate inhibition. Decreases in pH should inhibit heart LDH activity even in lower concentrations because, unlike skeletal tissue, aerobic tissues, such as those found in the heart muscles, do not sustain high buffering capacities (Castellini and Somero 1981).

Analyzing the effects of pH on the activities of LDH regarding substrate saturation, we can observe that the enzyme in the skeletal muscles assumes kinetics similar to those of the heart, i.e., high inhibition by low concentrations of pyruvate as the in vitro pH drops (Fig. 5.17). Thus, at least for *C. macropomum* and *M. duriventris*, LDH seems to be highly regulated, providing a better control of anaerobic glycolysis. However, a more complete study, including the study of temperature effects on the reverse side of the reaction, must be performed, i.e., conversion of lactate back to pyruvate when O_2 is available again. In addition, this study should include the glycolytic regulatory enzymes (PK and PFK).

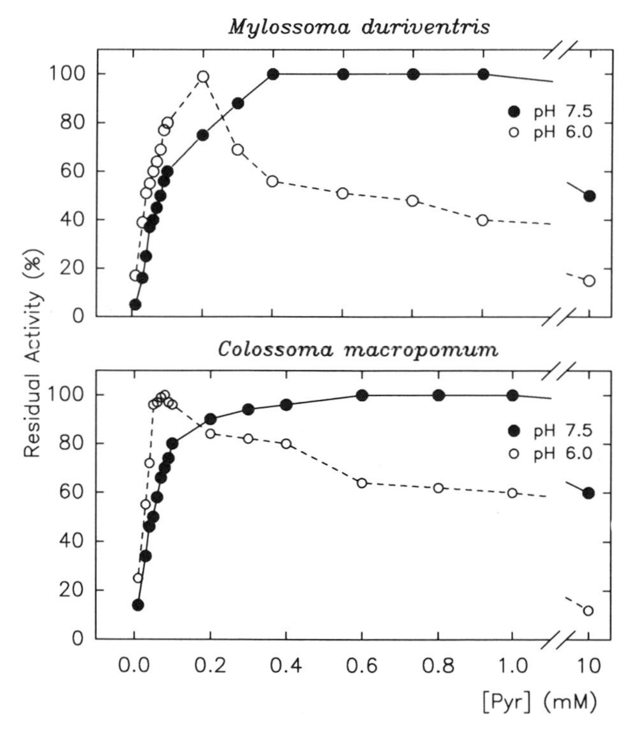

Fig. 5.17. LDH saturating curves (pyruvate reduction) of muscle from *M. duriventris* and *C. macropomum* as affected by low pH. Note that the enzyme assumes a heart-shaped curve at pH 6

5.3.3.1 Q_{10}

The term Q_{10} is defined as the change in the rate of any biological process (a catalytic rate, for example) over a 10 °C change in temperature. Metabolic rates of cold- and warm-acclimated ectotherms differ in their temperature responses. Cold-acclimated animals present larger Q_{10} values than do warm-acclimated ones (Hochachka and Guppy 1987). Furthermore, species acclimatized to wide ranges of temperatures have high Q_{10} values. Thus, tropical species are expected to exhibit lower Q_{10} values than temperate species.

Colossoma macropomum have very small Q_{10} values for LDH in both heart and skeletal muscle at both temperature intervals (Table 5.8). This behaviour is more emphasized for the LDH in the heart. Living in a thermally stable environment does not require enzymes to compensate for changes in the temperature. However, a complete thermoindependence is as extreme as having a very high Q_{10} such as that observed for the aquatic turtle (values close to 13 for blood respiratory properties; Herbert and Jackson 1985). *M. duriventris* usually exhibits Q_{10} values close to 2 (Table 5.8), demonstrating that thermoindependence is not a rule for the LDH of all tropical fish. The metabolic rate of *Colossoma macropomum* is affected by temperature (Saint-Paul 1983) as well as its blood properties (Val 1986). At high temperatures, this species decreases its O_2 consumption and Hb – O_2 affinities. This is an indication that oxidative metabolism is strongly affected and anaerobic glycolysis will also be activated. Small Q_{10} values prevent the hyperactivation of the enzyme which prevents lactate accumulation, particularly in the heart. It is most probable that the animal, at high temperatures, decreases its metabolic rate as well as its anaerobic metabolism. Depressing metabolism may be an alternative and low Q_{10}s in their enzyme reactions may be an adaptive characteristic.

Table 5.8. Q_{10} values for LDH activities from two Amazon serrasalmids. Two intervals of temperature (°C) are shown for three different pHs. All values were obtained in vitro. (Data from Almeida-Val 1986)

	pH	*M. duriventris*		*C. macropomum*	
		Skeletal	Heart	Skeletal	Heart
Q_{10}^{15-25}	6.0	1.9	1.0	0.6	1.1
Q_{10}^{25-35}		1.7	2.5	1.2	0.9
Q_{10}^{15-25}	7.5	1.5	1.3	0.8	1.4
Q_{10}^{25-35}		2.1	2.7	1.8	0.7
Q_{10}^{15-25}	8.5	2.4	1.9	0.4	1.2
Q_{10}^{25-35}		5.1	2.6	2.0	1.5

5.3.4 Thermal Acclimation

Thermal acclimation provides a better understanding of the responses exhibited by the metabolism in vivo under different temperatures. Temperature effects on physiological and biochemical properties have been studied mainly in temperate fish (Hazel and Prosser 1974; Graves and Somero 1982; Somero 1983; Coppes and Somero 1990). Compensatory responses during thermal acclimation occur through quantitative and qualitative changes in enzyme activities (Hochachka and Somero 1984). Thus, the acclimation process involves a reorganization of the animal's metabolic characteristics by scaling different rates for metabolic pathways, which is only possible by regulating enzyme activities (Hochachka and Somero 1984).

5.3.4.1 Electrophoretic Pattern

One of the first responses to thermal acclimation is a change in the amount of enzymes produced. This is possible by gene regulation. As *Colossoma macropomum* is unique in many of its respiratory strategies, and remains in waters with low O_2 levels, sometimes without developing the lips, it is one of the most-studied species from the adaptational point of view. This species, for example, differentiates synthesis of LDH-A subunits in heart during the acclimatization process (Fig. 5.1; see Almeida-Val et al. 1991b). These changes are thought to be qualitative and induced by gene regulation. Thus, in this case, small O_2 changes control the synthesis of one gene product, changing the aerobic/anaerobic balance of tissues. In the case of thermal acclimation, increasing temperature may induce an increase in the metabolic rate and, thus, in O_2 consumption. However, as we shall see next, this species presents a very unique response to temperature changes, when acclimated to different temperatures.

Many authors have suggested that isozymes play an important role in the thermal adaptation of ectotherms (Hochachka and Somero 1968, 1973; Yamawaki and Tsukuda 1979; Schwantes and Schwantes 1982), and that quantitative changes in isozyme activities are more important than qualitative changes during the acclimatization process (Somero 1975; Shaklee et al. 1977; Somero 1983). However, isozymic changes during the acclimation process are not so uncommon and are found in several enzymes in different fish species which were exposed to thermal acclimation. Some examples are lactate dehydrogenase (Hochachka 1965), pyruvate kinase (Somero and Hochachka 1968), acetyl cholinesterase (Baldwin and hochachka 1970), citrate synthase (Hochachka and Lewis 1971), isocitrate dehydrogenase (Moon and Hochachka 1971), alkaline phosphatase (Whitmore and Golberg 1972), and glucose-6-phosphate dehydrogenase (Hochachka and Clayton-Hochachka 1973). Most of these examples occur during cold acclimation in temperate fish.

Farias (1992), while studying thermal acclimation in the Amazonian fish species *Colossoma macropomum* and *Hoplosternum littorale*, reported that qualitative compensation occurs in *Colossoma macropomum* when acclimated to 20 °C. At this temperature the species shows changes in the electrophoretic pattern for LDH from its tissues (Farias 1992). Those changes were explained as the activation of a "silent" gene, tentatively termed as LDH-D* (Farias and Almeida-Val, in prep.). Because its presence is more accentuated in heart and retina, this new gene (probably a pseudogene) appears in order to compensate for the metabolism of aerobic organs. In addition, the product of this new LDH gene seems to copolymerize with the LDH-B* gene product (Fig. 5.18).

Thermal acclimation also induced changes in the LDH of goldfish (Hochachka 1965) most likely due to the activation of cold isozymes. The existence of cold and warm enzymes was first proposed by Fry and Hochachka (1970), and were explained as an alternative for metabolic compensation during thermal acclimation. Many authors have suggested that the presence of alternative isozymes in some animals is due to polymorphic loci, which may play a role in the thermal adaptation of the organism. Wilson et al. (1973), studying acclimated goldfish, suggested that the variation in the isozymic patterns was due to polymorphic loci instead of thermal isozymes. However, both trout and *Fundulus heteroclitus* were found to compensate thermally through changes in the LDH isozymic patterns (Bolaffi and Booke 1974).

Polymorphic LDH loci and polyploidy are not factors in the Amazonian *Colossoma macropomum*, according to previous studies (Almeida-Val et al. 1990; Almeida-Toledo et al. 1987, respectively). Furthermore, Almeida-

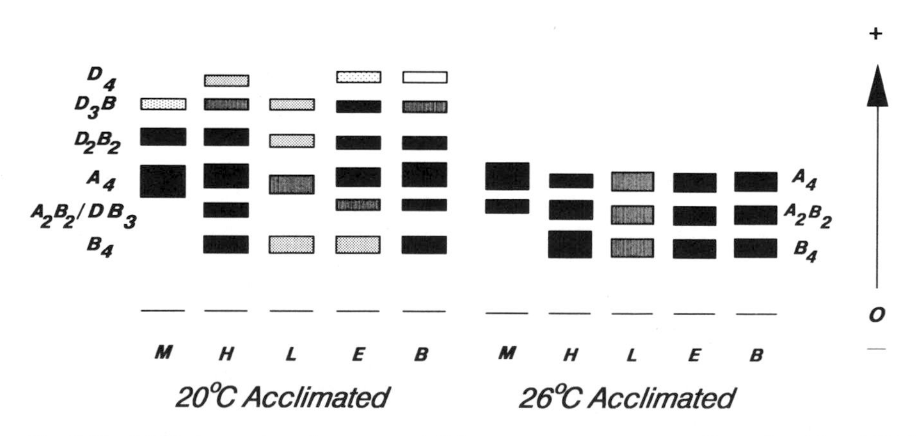

Fig. 5.18. LDH electrophoretic pattern of *Colossoma macropomum* acclimated at 26 and 20 °C. The electropherogram in fish acclimated to 20 °C indicates the appearance of new types of isozymes, suggesting the activation of a pseudogene. (Data from Farias and Almeida-Val, unpubl.) *M* Muscle; *H* heart; *L* liver; *E* eye; *B* brain

Toledo et al. (1987) reported a significant amount of redundant DNA in the karyotype of this species. Pseudogenes (i.e., genes that are made nonfunctional because they are no longer needed by the organism), which can be reactivated if the surrounding environment changes, are thought to exist in many organisms (Li 1983). As the temperature of the Amazon basin decreased to 18 or 20 °C during glacial periods (Bigarella and Ferreira 1985) and increased during interglacial periods, some specific gene loci may have lost their expression during periods in which they were no longer necessary. Because changes in temperature have occurred only on a geological basis in the Amazon region, activation of pseudogenes may represent a more economical solution for the organism than keeping an active gene even when it is not necessary. Animals who are acclimated to temperate regions regulate the synthesis of the LDH subunit in order to balance their metabolism at different temperatures. One of the best examples is the LDH adjustments that occur in the anaerobe aquatic turtle (Altman and Robin 1969). This regulation occurs on a yearly basis (spring/fall terms), when the gene is activated or deactivated for only short periods. This kind of regulation occurs more frequently in the Amazonian fish in response to frequent oscillations in the O_2 levels of their environment. The above-mentioned aspects help to explain why many authors propose that qualitative changes are limited to some environmental changes and that quantitative changes should be more important in the maintenance of the biochemical homeostasis of the organism (Somero 1983).

5.3.4.2 Km

Most enzymes never work at their maximum rate and substrate concentrations are always close to their Km values (reviewed by Hochachka and Somero 1973). However, the enzymes are able to work within a range that assures higher catalytic activities when substrate accumulates or when there are other regulatory and/or external stimuli. Thus, modulating Km values is a good alternative to control substrate oscillations that occur as a result of changes in the rates of metabolic pathways and/or directions. Because *Colossoma macropomum* and *Hoplosternum littorale* exhibit different responses to thermal acclimation processes, we consider it useful to assess each one separately.

Hoplosternum littorale is a silurid (Callichthyidae) that inhabits the bottom of lakes where O_2 contents are quite often very low. Several authors consider this species to be resistant to hypoxia and the environmental fluctuations of O_2 availability (Fink and Fink 1979; Affonso 1990). As mentioned earlier, this species is a facultative air-breather, using the intestine as the air-breathing organ. Although no qualitative changes are observed for the LDH patterns of this species, its responses to thermal acclimatization are unique (Farias 1992). First, *Hoplosternum littorale* are unable to survive exposure to 40 °C as do the *Colossoma macropomum* specimens. Secondly, at 30 °C, the Km values of LDH from skeletal muscle are lower than Km values of LDH from heart muscle. Due to peculiar properties of each isozyme, heart LDH always presents higher substrate affinities which are promptly inhibited when substrate levels increase. On the other hand, skeletal muscle isozymes are characterized by high Km values and noninhibition at saturating pyruvate concentrations. The reverse values observed for LDH Km values in *Hoplosternum littorale*, i.e., higher Km values for heart LDH, indicate that the heart is more suitable for anaerobic glycolysis under this condition. These characteristics are expected in a sedentary and anaerobe species. Actually, acclimatized animals (animals captured in nature) living at 30 °C also presented values similar to those obtained with acclimated animals (30 °C). At temperatures above and below 30 °C, LDH presents "normal" values, i.e., heart LDH has a lower Km compared to skeletal muscle LDH (Fig. 5.19; see Farias 1992).

Although thermal compensatory responses are important for the homeostasis of the enzyme, some authors suggest that stable enzyme Km would help to reduce stress of the organisms caused by changes in the temperature of their habitat (reviewed by Hochachka and Somero 1984). The differences among the values obtained for heart and skeletal muscle LDH of *Colossoma macropomum* at all temperatures are small, suggesting that the LDH Km values are thermoindependent.

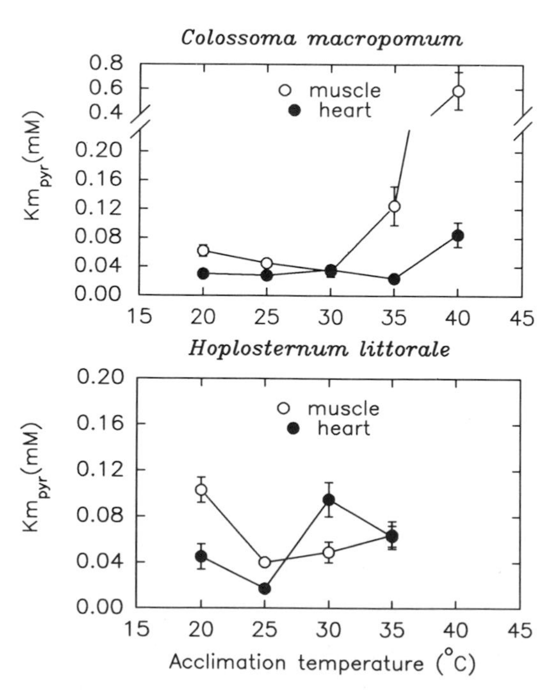

Fig. 5.19. LDH $Km_{(pyr)}$ values for *Colossoma macropomum* and *Hoplosternum littorale* acclimated at different temperatures. Data from Farias and Almeida-Val (unpubl.)

Studying LDH from skeletal muscle (LDH-A_4) of fish species living in different thermal regimes, Coppes and Somero (1990) suggested that species facing a large temperature range (eurytherm) keep their Km values for pyruvate constant, while species facing a narrow temperature range (stenotherm) exhibit compensatory responses to thermal acclimation. Lactate dehydrogenase Km values obtained for *Hoplosternun littorale* are very low compared to other teleost muscles. In addition, the Km values did not change according to thermal acclimation as expected in a stenotherm species. These reduced changes in LDH Km during the thermal acclimation reflect reduced changes in the metabolism of the animal. In fact, during the thermal acclimation process, *Hoplosternum littorale* presents different respiratory responses, i.e., at low temperatures the animal remains close to the bottom of the aquarium and relies preferentially on water-breathing, increasing the air-breathing frequency when exposed to high temperatures.

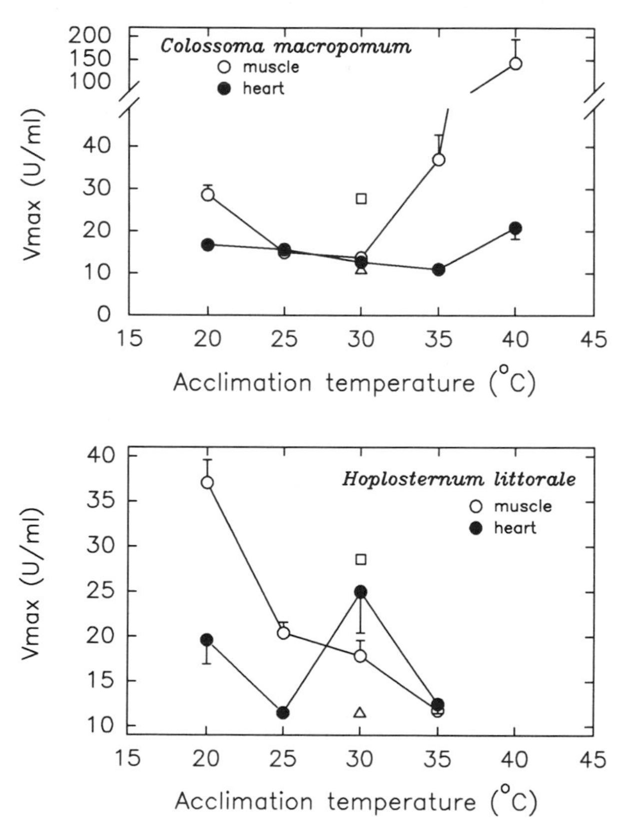

Fig. 5.20. V_{max} values for *Colossoma macropomum* and *Hoplosternum littorale* acclimated at different temperatures. Values for muscle LDH (*squares*) of wild animals are higher than acclimated animals at a similar temperature. The inverse situation is observed for heart LDH (*triangles*)

Thus, the compensatory responses to thermal acclimation must be scaled according to the energy turnover changes and then according to different O_2 requirements. Due to these characteristics, it seems possible that LDH activities decrease when the animal is exposed to increasing acclimation temperatures, mainly in skeletal muscle (Fig. 5.20). At low temperatures the organism relies primarily on water-breathing and must balance its energy requirements with anaerobic glycolysis in skeletal muscle. When the temperature increases, the drop in the water dissolved O_2 must be the signal to activate air-breathing. After activating air-breathing, O_2 becomes available to all tissues and anaerobic glycolysis rates are decreased.

The development of the specialized lower lip in *Colossoma macropomum* during low O_2 conditions provides increased O_2 uptake from the water's surface, contributing to a decreased lactate accumulation in the tissues (Almeida-Val et al. 1993). However, it is not clear whether this adaptation will restore the balance between anaerobic and aerobic metabolism in vital organs such as the heart and brain. The acclimatization at 40 °C lead the animal to anaerobic metabolism because all specimens exhibited ASR and developed lips. At this temperature both Vmax and Km increase for the LDH in both skeletal and heart muscles. On the other hand, no significant changes were observed in Km values for the LDH in both heart and skeletal muscles at temperatures below 40 °C. Similar results were found when this enzyme was submitted to different temperatures in vitro (Almeida-Val et al. 1991b). Thus, except for the temperature of 40 °C, this species exhibits high homeostasis concerning LDH functions and, consequently, anaerobic glycolysis. The low Km values for the LDH in skeletal muscle have also been observed in *M. duriventris* (Almeida-Val et al. 1991b) and in *H. littorale* (Farias 1992). These findings support the idea of the selection of heart-type subunits for LDH (Almeida-Val et al. 1993). The environmental characteristics of the Amazon, which impose chronic hypoxia on the aquatic environment, induce the organism to sustain a low energy demand with no anaerobic activation, unless energy is absolutely necessary. A heart-type LDH would facilitate this metabolic balance as seen in *C. macropomum*.

Because most enzymes function in undersaturated conditions, the temperature effects on their Km have an effect on their reaction velocities. Indeed, no changes in the maximum velocities for pyruvate-lactate conversion in muscle of *C. macropomum* have been observed during acclimatization (Almeida-Val et al. 1991b). The presence of such adaptive features in species living in a narrow range of temperatures is an indication that the acclimation and/or acclimatization processes cannot be estimated based on the species habits as suggested by Coppes and Somero (1990).

5.3.4.3 *Regulating Q_{10} Values*

High Q_{10} values are characteristic of cold-acclimated ectotherms which helps to decrease metabolic rate when O_2 is depleted in the aquatic environment or when the organism becomes torpid, depressing energy turnover rates. This mechanism is called inverse temperature compensation and has been demonstrated in fish, reptiles, and amphibians (reviewed by Hochachka and Somero 1984). All these examples occur in temperate ectotherms, which must cope with high temperature oscillations and large and long-term temperature drops. High Q_{10} values for metabolic rate are found in the aquatic turtle, which depresses its metabolism and is able to hold its breath for up to 6 months during winter (Herbert and Jackson 1985; Almeida-Val et al. 1994). In this species, each different type of tissue has its own Q_{10} value, thus, each tissue responds differently to drops in temperature. It follows, then, that the Q_{10} values differ according to the studied organ and to the studied process (Almeida-Val et al. 1994). The aquatic turtle also offers a good example of metabolic depression of both oxidative and anaerobic glycolysis pathways while overwintering on the bottom of ice-covered lakes (Hochachka 1980).

Tropical fish are able to maintain constant metabolic rates mainly because they have stable thermal conditions. Upon reviewing the subject, Hochachka and Guppy (1987) suggested that warm-acclimated animals possess low Q_{10} values. The LDH in *Colossoma macropomum* is characterized by low Q_{10} values in both skeletal and heart muscles. The Q_{10} values for these species increase only at 40 °C. According to Somero (1969), Q_{10} values below 1 are characteristic of sedentary species that are not able to sustain high metabolic activities. In conclusion, if acclimatization to low O_2 concentrations induces the organism to sustain a low metabolic rate, and if small oscillations in temperature do not affect its metabolism, then the animal will be able to perform well when exposed to chronic hypoxic environments such as those found in the Amazon.

Similar characteristics have been described for *H. littorale*, with the exception of the heart muscle at 25–30 °C (Farias 1992). Values lower than 1 are typical in both species, showing that LDH activation occurs at low temperatures. Again, this could be seen as adjustments in response to factors other than temperature, such as O_2, pH, or river water level. Thus, it is important to verify the effects of temperature on the overall metabolism in order to observe the responses to external pressures.

In conclusion, the effects of temperature changes in the metabolism of Amazonian fish should be considered simultaneously with the biological characteristics of the species and other environmental parameters. Almost all Amazonian fish possess some kind of adaptation to survive the simultaneous high temperatures and chronic hypoxia. The role of LDH at the end of anaerobic glycolysis makes this enzyme an excellent tool for the analysis of environmental variations. The increases in the LDH activities observed in the heart and muscles of *C. macropomum* and *H. littorale* during in vivo and in vitro exposure to low temperatures (Almeida-Val et al. 1991b; Farias 1992) suggest that it is necessary to activate the anaerobic metabolism i.e., a temperature drop causes some alteration in the tissular/cellular O_2 availability (Val 1986; Affonso 1990).

5.4 Metabolic Arrest

For temperate aquatic ectotherms, living without O_2 occurs periodically, mainly during the cold season. Thus, changes in environmental temperature presents an excellent tool for depressing metabolism (Hochachka and Guppy 1987). For tropical aquatic ectotherms, high water temperatures and high rates of O_2 utilization may lead to severe molecular O_2 limitations. In this situation, the organisms cannot use the environmental temperature to help depress its metabolism. Again, by depressing its metabolism and achieving metabolic arrest, the aquatic turtle sustains anoxia for more than 24 h while diving during summer temperatures, i.e, at temperatures greater than 20 °C (Hochachka 1980; Hochachka and Guppy 1987). For tropical fish, there are only two options available: sustaining low oxidative rates and depressing metabolism when it is necessary and/or sustaining high anaerobic rates in their tissues. Nevertheless, Amazonian fish can survive anoxic situations; O_2 depletion does not exist as a long-term pressure and, thus, we do not expect to find examples of animals, like the aquatic turtle, which can survive long-term anoxia, in the Amazon. However, anoxia tolerance is present in the water-breathers *Hoplias malabaricus* (Driedzic et al. 1978) as well as in *Astronotus ocellatus* (Soares 1993; pers. observ.), species with no anatomical or morphological respiratory adaptations. It is possible that both species slow their activities down in order to survive anoxic conditions (Soares 1993; pers. observ.). Thus, as in the aquatic turtle, the environmental signal for metabolic depression seems to be the lack of oxygen.

As stated above, the only reported example of metabolic arrest in Amazon fishes occurs in the lungfish (Hochachka 1988b). Metabolic changes during lungfish diving have been reported by Dunn et al. (1983) and this species maintains its energy balance during the whole diving period. Activation of anaerobic glycolysis to compensate for drops in ATP levels has not been reported (Hochachka 1993). Also, this animal undergoes several months of metabolic depression during periods of drought and it makes many metabolic adjustments in order to sustain the estivation period (reviewed by Hochachka and Guppy 1987). It is therefore worthwhile to present an overall idea of what those mechanisms are and how the fish can cope with the absence of water at high temperatures.

5.4.1 Lungfish Estivation

Most estivating animals possess common features; they can sustain very low metabolic rates (as previously discussed), hypophagia and inactivity, and biochemical adjustments in their different organs protect against the lack of water as well as other factors (Hochachka and Guppy 1987). As mentioned above, adjustments of enzymes occur in different ways in different organs. All these adjustments lead to a single point: an overall low metabolic rate.

All tissues in the lungfish store high amounts of glycogen, which indicates the high glycolytic potential (for both aerobic and anaerobic pathways). The utilization of glucose, even in oxidative pathways, has been considered an adaptive characteristic because it increases the ATP yield and avoids switching on and off different metabolic pathways (Hochachka 1993). Other than glucose, the Amazon lungfish also utilizes fat and its heart can be considered a high oxidative organ, compared with other organs. However, compared with other fish, the heart and brain of lungfish display higher glycolytic but lower oxidative potentials (reviewed by Almeida-Val and Hochachka, in press). High gluconeogenesis capacities are found in its liver and kidney. Hochachka (1980) found high rates of amino acid metabolism and urea cycle in those tissues as well. Several reports agree that lungfish retain a multisubstrate-based metabolism and are able to utilize proteins and amino acids as fuels in addition to the ones just listed (reviewed by Hochachka and Guppy 1987). The utilization of protein and amino acids may be necessary during estivation and has been confirmed since Homer Smith first studied this process in 1930. He recognized the utilization of multiple substrates at the beginning of the lungfish estivation process. No further studies have been done on protein and amino acid metabolization during estivation of lungfish. Glycogen stores are not completely depleted during estivation periods. This indicates that the lungfish metabolism relies on more than one type of substrate. Fasting fishes use protein and amino acid as substrate for metabolism (Mommsem et al. 1980). Once the proteins or amino acids are no longer available from the blood of estivating lungfish, the protein stored in the muscle must be mobilized. Alanine is perhaps oxidized in most tissues except in the liver and kidney, where its main function is gluconeogenesis (Hochachka and Guppy 1987).

The white muscles in lungfish store glycogen as large-diameter α-particles or glycogen rosettes. This amount of glycogen, and the way it is stored, are unusual in white muscles. White muscles in fish do not have large amounts of glycogen (Pritchard et al. 1974) and the only other example is tuna (Hochachka et al. 1979). The presence of glycogen in the red muscles of fish is common. However, this characteristic is exaggerated in the lungfish because the number of glycogen granules is so excessive that Hochachka and Hulbert referred to them as glycogen "seas". Glycogen bodies form the glycogen "seas" which are not located in all parts of the cell in the red muscle; it appears regularly and usually packed around the nuclei (Hochachka and Hulbert 1978). The heart tissue retains an interfibrillar glycogen body formation that is associated with an unusual type of sarcoplasmic reticulum. These glycogen deposits suggest that lungfish preferentially uses glycogen for both aerobic and anaerobic metabolism.

However, Hochachka and Guppy (1987) assert that this amount of stored glycogen is not adequate to sustain long-term estivation. As mentioned earlier, protein catabolism is performed during lungfish estivation and glycogen stores are kept during this period. Glycogen and glucose are thought to be reserved for those cells that have an exclusive requirement for glucose (Hochachka 1980).

Compartmentalization of glycogen also occurs in *Synbranchus marmoratus*, another air-breather. This species displays a similar metabolic organization as lungfish, exhibiting estivation habits as well. *Osteoglossum bicirrhosum*, a water-breather, also retains high quantities of glycogen granules connected with sarcoplasmic reticulum in its heart. These glycogen granules are intercalated with regions containing many mitochondria (Hochachka et al. 1978b).

The evolution of mechanisms protecting the organisms against hypoxia clearly follows some set of "rules". Fish adaptation patterns must be similar, especially when they must cope with similar environmental constraints. Selecting glucose metabolism seems to be a widespread strategy that occurs, to various degrees, in organisms ranging from fish to man (P.W. Hochachka, pers. comm.). Sustaining low oxidative rates is another strategy which occurs in Amazonia fish in general.

Depressing metabolism (metabolic arrest and reverse Pasteur effect) has been described as a universal response to many environmental constraints such as O_2 deficits, large temperature drops, dry periods, and so on. Oxygen depletion can last for short-term periods (as occurs during diving) or long-term periods. In the Amazon, short anoxic periods occur at night or over 1 or 2 days at most. The presence of a reverse Pasteur effect and depression of oxidative metabolism are probably common in Amazon fishes as well.

5.5 Effects of Specific Environmental Conditions

Sudden natural modifications to the environment induce constraints on the organisms which are very often unable to rapidly adjust their metabolic machinery. Similarly, these organisms are unable to buffer modifications caused by the activities of men. In the Amazon, these environmental modifications include the sudden increase in the dissolved H_2S concentrations in the water during the *friagem* period, water contamination with petroleum and mercury, and the release of anoxic water downstream of the hydroelectric power plants. Most of these modifications force the organisms to adjust their metabolism in order to maintain an aerobic/anaerobic balance.

5.5.1 Accumulating End Products

Avoiding self-pollution is one of the most important adaptations against the effects of hypoxia. The accumulation of end products such as lactate in aerobic tissues without damage occurs only in animals that are good anaerobes (Hochachka and Somero 1984). We have reviewed how many fish deal with lactate accumulation and how their LDH isozyme systems have evolved with a predominance of heart-type LDHs which have their function inhibited by large quantities of pyruvate (Hochachka and Hulbert 1978; Almeida-Val et al. 1993; Farias 1992).

5.5.1.1 Lactate Accumulation

One of the main problems of relying on anaerobic metabolism during periods of O_2 depletion is the accumulation of acidic end products in sensitive organs, such as the heart. In fish, the solutions to this problem are not common and always involve a depression of anaerobic glycolysis, or the rapid "washing out" of lactate which has accumulated mainly in the blood. Selection of specific LDH isozymes and their differential distribution in different organs help the organism to buffer the undesirable effects. The unique solution of producing ethanol and CO_2 as end products is present in the cyprinid *Carassius auratus*, the goldfish. This fish is considered a good anaerobe, comparable to the aquatic turtle. It is able to sustain long periods of anoxia while overwintering in ice-covered lakes. The production of ethanol and CO_2 was first discovered by Shoubridge and Hochachka (1980) and, since then, the mechanisms have been clarified by many other authors. In a recent report, Van Waarde et al. (1993) reported the responses of the goldfish, regarding the activation of anaerobic glycolysis and the consequent accumulation of lactate in its tissues, to anoxia exposure. Subsequent drops in pH activate pyruvate decarboxylase, channelling pyruvate to ethanol via acetaldehyde. After that, a steady state is thought to develop where lactate is removed at the same rate as it is formed. Thus, anaerobic glycolysis seems to be the first response to limiting O_2 supplies in goldfish as well as in other species, but then different strategies are used to place a limit on further lactate accumulation. In goldfish, ethanol production is probably coupled with some energy-demand suppression.

In the same report, Van Waarde et al. (1993) review the literature regarding the presence of this mechanism in different fish species. The only two other species exhibiting the same strategy belong to cyprinids. In their checklist, at least one important Amazon fish species is included *Colossoma macropomum*. Besides surviving under extreme hypoxic conditions, this species has a different solution for lactate accumulation which involves some morphological and behavioural adaptations (reviewed by Almeida-Val et al. 1993).

The presence of a similar strategy in Amazon fishes cannot be discarded since some species have a similar tolerance levels to anoxia as the goldfish. As already mentioned, at least two Amazonian species are able to survive anoxia for more than 5 h by depressing their metabolism: *Hoplias malabaricus* (Erythrinidae) and *Astronotus ocellatus* (Cichlidae) (Soares 1993; pers. observ.). According to Shoubridge (1980), ethanol in goldfish is produced in the muscle by transforming acetaldehyde into CO_2 and ethanol via alcohol dehydrogenase (ADH). Most fish have ADH only in liver tissues. Shoubridge (1980) demonstrated that the ADH in the liver and muscle of goldfish is encoded at different loci, as indicated by the presence of different isozymes which are detected electrophoretically.

Our studies on different cichlid species, including *Astronotus ocellatus*, revealed the presence of different isozymes for LDH in both muscle and heart of some species (Paula-Silva and Almeida-Val, unpubl. data). As ADH can appear during the histochemical staining of different dehydrogenases, the presence of extra bands could be due to ADH, in addition to LDH. Further studies on cichlids must be developed, because this group contains good anaerobes and displays no evident behavioural, morphological, or anatomical strategy to survive hypoxia or anoxia during long-term exposure.

5.5.2.2 Ammonia Accumulation

Another problem is the nitrogen metabolism which, in most fish, ends with the production of ammonia, a highly toxic compound that, due to its excretion in large amounts of water, does not represent a polluting agent for most species. However, for estivating organisms, mainly those that burrow in the mud (for example, lungfish), the use of ammonia as an end product may cause self-pollution. In addition, as reviewed above, lungfish utilize protein during estivation as a substrate for oxidative pathways. Utilizing nitrogen compounds (for example, amino acids and proteins) is a good adaptive characteristic because food is not available during estivation process, but one problem arises: how to avoid or cope with the end products which accumulate, especially toxic ones like ammonia.

One of the solutions is the production of a less toxic end product. This seems to be the solution adopted by lungfish which synthesize another less toxic product: urea, the most common excretion product in higher vertebrates.

The ability of lungfishes to produce urea during protein metabolism makes it one of the best examples of the transition from aquatic to terrestrial life (Hochachka and Somero 1973). One of the first processes to moderate when the lungfish starts to estivate is ammonia production. However, the rates of urea formation are maintained. Based on other studies and current information on fish physiology, the rate at which urea is formed by the lungfish should be intolerable to the fish. Hochachka and Guppy (1987) hypothesized that the lungfish is able to recycle the urea in some way during estivation. In addition, because the Amazonian lungfish does not form a cocoon around itself during estivation, as do the African lungfish, the problem of self-pollution is most likely reduced due to the open system (Hochachka 1980).

Chapter 6

Conclusion and Perspectives

The preceding chapters in this book dealt with environmental characteristics of the Amazon basin, its distinct ichthyofauna, and the physiological and biochemical characteristics of the respiration of representative fish species of this basin. In addition, we have tried to analyze the way in which the fishes of the Amazon deal with the "pulsative" environment. Here, general conclusions on the subject are presented. However, as mentioned earlier, the nature of the interactions of the fish with the environment, the dynamics of flooding cycles, the extension of habitat overlaps and intergradations, the intraspecific diversity, and the endless number of solutions adopted by the fish to survive environmental constraints, restrict our conclusions and multiply perspectives.

Surviving the Amazon constraints has induced a myriad of solutions. The exploration of new habitats has imposed, to some extent, the existence of biological alternatives. As a consequence, a high speciation rate occurs, as indicated by the number of fish species and intraspecific variability. The environmental oscillation, in turn, has demanded plasticity in the regulation of the metabolic machinery. Extensive protein variability has provided this necessary plasticity. In addition, this protein heterogeneity, hidden in the nucleus of each cell, has supported much of the evolutionary experimentation. In fact, adaptive strategies are observed at all biological levels.

The opportunistic character of the feeding behaviour, i.e., the capacity to feed on the available food, has played an important role in the process of colonization of ever-changing habitats. Migration cycles are also influenced by environmental oscillations. The fries and alevins are dependent on adjustments of spawning migration to access nutrient-rich areas. After spawning, the spent animals migrate back and disperse across the flooded forest. As the water level begins to fall, the animals congregate and migrate upstream again.

Oxygen availability is another important factor shaping the relationship between fish and the environment. Different ways to maximize O_2 transfer have been achieved. The utilization of air as a source of O_2 has been attempted independently many times during the explosive radiation of fish. At least nine families of Amazonian fish have air-breathing representatives. Some are obligatory air-breathers, others are facultative. Modifications of the stomach and intestine as in catfish, the vascularized swim bladder as in

Arapaima and *Hoplerythrinus*, the diverticulated and vascularized mouth as in *Electrophorus*, and the lungs as in *Lepidosiren*, are examples of air-breathing structures. Many strategies have also been adopted by gill breathers to improve O_2 transfer from poorly oxygenated waters. They include skimming the waters surface, optimization of O_2 uptake, and even adjustments of the metabolic machinery to the current O_2 avilability. Several fish species, however, are able to detect O_2 depletions in their environment or even to anticipate them, leaving an area before large drops in the water O_2 content occur.

Adjustment of blood characteristics have been shown for many fish species of the Amazon. These adjustments are clearly induced by environmental changes. The number of circulating erythrocytes, intraerythrocytic phosphate levels, and type of hemoglobins, for example, are all related to water O_2 content and river water level. The main goal of these adjustments is to optimize O_2 transfer.

The capacity to tolerate hypoxia or even anoxia has been observed in many fish species of the Amazon. This capacity is related to biochemical adjustments of the metabolic machinery. Many organisms are able to ensure O_2 delivery to highly aerobic tissues (heart and brain) and/or to shut down the energy demand of some metabolic processes. This is achieved by regulation of enzyme activities. This regulation is continuous because O_2 availability changes seasonally or even daily. Tissue LDH distribution is illustrative of this aspect.

Paradoxically, air-breathers, with a potentially unlimited O_2 supply, also sustain low metabolic rates and high anaerobic potential. This is necessary in order to reduce energy demand during diving periods. In addition, air-breathing in fish was not developed to achieve a more efficient energy turnover and, consequently, not to improve endurance. It is most probable that air-breathing habits in fish were developed only to avoid the constraints imposed by low O_2 concentrations in the aquatic environment.

6.1 Perspectives

The Amazon is a biological "gold mine". The hiatus in our understanding of the relationship between the fishes of the Amazon and their environment is impressive. The unknown far exceeds the known. In reviewing the material for the previous chapters and from the discussions with many colleagues, we have selected some issues for further studies. They are summarized as follows:

1. The fries and alevins are passively carried to *várzea* lakes where poor O_2 conditions are often observed. It is important to analyze the strategies developed to deal with such conditions during the first stages of life and to realize how these conditions affect the development of the animals.

2. It has been shown that active species of O_2 (O_2 radicals) are generated during exposure to hypoxia. These radicals affect many metabolic processes. An analysis of the protective mechanisms developed by fish of the Amazon will enlarge our understanding of the prevention of cell damage.
3. Morphological adaptations (lip development, vascularization of stomach and intestine) are observed in many fish species of the Amazon. The extent to which they improve O_2 uptake and their effectiveness under different environmental conditions need to be investigated.
4. Morphological adaptations and metabolic adjustments occur simultaneously in several fish species. The extent to which one affects the other is unknown and needs to be analyzed.
5. Long migration is characteristic of many fish species of the Amazon. During migration, fishes are exposed to different physiological and environmental conditions. A comparative analysis of their endurance capacity and metabolic preferences will indicate the strategies utilized to sustain energy turnover rates during exercise under tropical conditions.
6. Fish schools are exposed to low O_2 conditions during migration. Field observations suggest that specific movements of the fish may increase water aeration. This could be an important strategy to sustain the metabolic rates in migratory tropical fishes and needs to be studied.
7. The fishes of the Amazon have been exposed to regular environmental fluctuations. Thus, this group is unique for studies of stress responses.
8. It has been shown that CO_2 excretion and O_2 uptake are coupled in fish. Two questions arise from the analyzis of fishes of the Amazon: how do facultative air-breathers deal with CO_2 excretion during transition from air- to water-breathing? How can air-breathers take up O_2 in one place (the lung) and excrete CO_2 in a different place (the gills)?
9. Multiple adaptive responses to low O_2 conditions have been described in fishes of the Amazon. Some of them are activated either to improve the O_2 uptake or to reduce the O_2 demand. In case these fail, other mechanisms are adjusted to protect the animal. The way in which these adaptive responses are simultaneously activated in fish of the Amazon is unknown and needs to be investigated.
10. A significant range of water pH is observed in the Amazon. However, the effect of pH on both O_2 transfer and metabolic processes in fishes of the Amazon is poorly known and requires further studies.
11. Much of the metabolic machinery is designed to meet dietary constraints. The opportunistic character of the feeding behaviour of fishes of the Amazon raises a new group of questions. How are the enzymes regulated, distributed, and qualitatively altered during changes in feeding behaviour?
12. "High performance fishes" of the Amazon are also "fast growing fish" – pirarucu is outstanding – as are other species. How is energy versus carbon preferentially partitioned for growth, protein biosynthesis, and

differentiation? These are unknown problems that need to be examined. An especially crucial problem arises if the demands of high growth conflict with hypoxia tolerance. In other species, protein synthesis is "supersensitive" to hypoxia; what happens to fast growing Amazon fishes when they encounter hypoxic conditions? This is also unknown, and is our final example illustrating why the fishes of the Amazon are an underexplored biological "gold mine".

References

Ab'Saber AN (1977) Espaços ocupados pela expansão dos climas secos na América do Sul, por ocasião dos períodos glaciais quaternários. Paleoclimas 3:1–18

Ab'Sáber AN (1979) Os mecanismos de desintegração das paisagens tropicais no pleistoceno. Efeitos paleoclimáticos do período Wüm-Wisconsin no Brasil. Inter-facies 4:1–19

Ab'Sáber AN (1982) The paleoclimate and paleoecology of Brazilian Amazonia. In: Prance GT (ed) Biological diversification in the Tropics. Columbia Univ Press, New York, pp 74–77

Absy ML, Cleef AM, Fournier M, Martin L, Servant M, Sifeddine A, Silva FFM, Soubies F, Suguio K, Turcq B, Vam der Hammem T (1991) Mise en évidence de quatre phases d'overture de la forêt dense dans le sud-est de l'Amazonie au cours des 60 000 derniéres années. Première comparison avec d'autres régions tropicales. C R Acad Sci Paris 312(II): 673–678

Achaval ZC (1986) Isozymes of glucose phosphate isomerase (PGI) in fishes of the subclass Actinopterigii. Comp Biochem Physiol 84B(4):575–588

Adis J (1984) Seasonal igapó-forests of central Amazonian black-water rivers and their terrestrial arthropod fauna. In: Sioli H (ed) The Amazon. Limnology and landscape ecology of a mighty tropical river and its basin. Junk, Dordrecht, pp 245–268

Affonso EG (1990) Estudo sazonal de características respiratórias do sangue de *Hoplosternum littorale* (Siluriformes, Callichthyidae) da ilha da Marchantaria, Amazonas. MSc Thesis, PPG INPA/FUA, Manaus, 87 pp

Ali MA, Anctil M (1976) Retinas of fishes. Springer, Berlin Heidelberg New York, 284 pp

Ali MA, Klyne M (1985) Vision in vertebrates. Plenum Press, New York, 272 pp

Almeida RG (1984) Biologia alimentar de três espécies de *Triportheus* (Pisces, Characoidei, Characidae) do lago Castanho, Amazonas. Acta Amazonica 14(1/2):48–76

Almeida-Toledo LF, Foresti F, Toledo-Filho SA (1985) Spontaneous triploidy and NOR activity in *Eigenmannia* sp. (Pisces, Sternopygidae) from the Amazon basin. Genetica 66:85–88

Almeida-Toledo LF, Foresti F, Toledo-Filho SA, Bernardino G, Ferrari W, Alcântara RCG (1987) Cytogenetics study on *Colossoma mitrei*, *Colossoma macropomum* and their interspecific hybrid. Proc World Symp on Selection, hybridization, and genetic engineering in aquaculture, Vol. 1. Bordeaux 27–30 May, 1986, Berlin 1987, pp 189–195

Almeida-Val VMF (1986) Lactato desidrogenase de duas espécies de peixes da Amazônia, *Mylossoma duriventris* e *Colossoma macropomum* (Characiformes): aspectos adaptativos. PhD Thesis, PPG INPA/FUA, Manaus, 141 pp

Almeida-Val VMF, Hochachka PW (1994) Air-breathing fishes: metabolic biochemistry of the first diving vertebrates. In: Hochachka PW, Mommsen TP (eds) The biochemistry and molecular biology of fishes: environmental and ecological biochemistry, vol 5. Elsevier, Amsterdam (in press)

Almeida-Val VMF, Val AL (1990) Adaptação bioquímica em peixes da Amazônia. Ciênc Hoje 11(64):62–67

Almeida-Val VMF, Val AL (1993) Evolutionary trends of LDH isozymes in fishes. Comp Biochem Physiol 105B:21–28

Almeida-Val VMF, Schwantes AR, Val AL (1985) Electrophoretic patterns of hemoglobin and oxygen binding properties of blood of Anostomidae fishes from Parana-Pardo-Grande hydrographyc basin (São Paulo state, Brazil). J Exp Zool 235:21–26

Almeida-Val VMF, Schwantes ML, Val AL (1990) LDH isozymes in Amazon fish. I. Electrophoretic studies on two species from Serrasalmidae family: *Mylossoma duriventris* and *Colossoma macropomum*. Comp Biochem Physiol 95B:77–84

Almeida-Val VMF, Schwantes ML, Val AL (1991a) LDH isozymes in Amazon fish-II. Temperature effects in LDH kinetic properties from *Mylossoma duriventris* and *Colossoma macropomum* (Serrasalmidae). Comp Biochem Physiol 98B:79–86

Almeida-Val VMF, Val AL, Feldberg E, Caraciolo MCM, Porto JIR (1991b) Evolução de peixes da Amazônia: aspectos genéticos e adaptativos. In: Val AL, Feldberg E, Figliuolo R (eds) Bases cientificas para estratégias de preservação e desenvolvimento da Amazônia: fatos e perspectivas. Imprensa da Universidade do Amazonas, Manaus, pp 281–298

Almeida-Val VMF, Paula-Silva MN, Caraciolo MCM, Mesquita LSB, Farias IP, Val AL (1992) LDH isozymes in Amazon fish. III. Distribution patterns and functional properties in Serrasalmidae (Teleostei: Ostariophysi). Comp Biochem Physiol 103B:119–125

Almeida-Val VMF, Val AL, Hochachka PW (1993) Hypoxia tolerance in Amazon fishes: status of an under-explored biological "goldmine". In: Hochachka PW, Van den Thillart G, Lutz P (eds) Surviving hypoxia: mechanisms of control versus adaptation. CRC Press, Boca Raton (in press)

Almeida-Val VMF, Buck LT, Hochachka PW (1994) Substrate and temperature effects on turtle heart and liver mitochondria. Am J Physiol 266:R858–R862

Altman M, Robin ED (1969) Survival during prolonged anaerobiosis as a function of unusual adaptation involving lactate dehydrogenase subunits. Comp Biochem Physiol 30:1179–1187

Amadio SA (1986) Estudos de ecologia e controle ambiental na região do reservatório da UHE Balbina. Relatório do subprojeto "Levantamento ictiofaunístico". Convênio Eletronorte/CNPq-INPA, Manaus, 78 pp

Araújo-Lima CARM (1984) Distribuição espacial e temporal de larvas de Characiformes em um setor do rio Solimões-Amazonas, próximo a Manaus, AM. MSc Thesis, PPG INPA/FUA, Manaus, 86 pp

Araújo-Lima CARM (1991) Patterns of development of larval fish of Central Amazon. Eur Aquac Soc Publ 15:271–273

Araújo-Lima CARM, Forsberg B, Victoria R, Martinelli L (1986) Energy Sources for detritivorous fishes in the Amazon. Science 234:1256–1258

Aragão LP (1981) Ecologia, desenvolvimento ontogenético, alimentação e reprodução do aruanã, *Osteoglossum bicirrhosum* Vandelli, 1829, do lago Castanho, Amazonas. MSc Thesis, PPG INPA/FUA, Manaus, 78 pp

Bailey PB (1981) Fish yield from the Amazon in Brazil. Comparison with African river yields and management possibilities. Trans Am Fish Soc 100:351–359

Bailey PB (1982) Central Amazon fish populations: biomass, production and some dynamic characteristics. PhD Thesis, Dalhousie Univ, Halifax, 308 pp

Baker HC (1970) Evolution in the tropics. Biotropica 2(2):101–111

Baker RR (1978) The evolutionary ecology of animal migration. Holmes & Meir, New York, 1012 pp

Baldwin J, Hochachka PW (1970) Functional significane of isozymes in thermal acclimation. Biochem J 116:883–887

Bartlett GR (1976) Iron nucleotides in human and rat red cells. Biochem Biophys Res Commun 70:1063–1070

Bartlett GR (1978a) Phosphates in red cells of two lungfish: the South American, *Lepidosiren paradoxa*, and the African, *Protopterus aethiopicus*. Can J Zool 56:882–886

Bartlett GR (1978b) Phosphates in red cells of two South American osteoglossids: *Arapaima gigas* and *Osteoglossum bicirrhosum*. Can J Zool 56:878–881

Bartlett, GR (1978c) Phosphate compounds in reptilian and avian red blood cells: developmental changes. Comp Biochem Physiol 61A:191–202

Bartlett GR (1980) Phosphate compounds in vertebrate red blood cells. Am Zool 20:103–114
Bartlett GR, Borgese TA (1976) Phosphate compounds in red cells of the chicken and duck embryo and hatchling. Comp Biochem Physiol 55:207–210
Bartlett GR, Schwantes AR, Val AL (1987) Studies on the influence of nitrite on methemoglobin formation in Amazonian fishes. Comp Biochem Physiol 86C:449–456
Beall RJ, Privitera CA (1973) Effects of cold exposure on cardiac metabolism of the turtle *Pseudemys (Chrysemis) picta*. Am J Physiol 224(2):435–441
Beebe W (1945) Vertebrate fauna of a tropical dry season mud-hole. Zoologica 30:81–87
Belokopytin YS, Rakitskaya LV (1981) Hematological indices of ecologically different sea fishes, the black sea scad *Trachurus mediterraneus ponticus* and the black sea surmullet, *Mullus barbatus ponticus* at rest and under muscular load. J Ichthyol 21:69–76
Berg T, Steen JB (1965) Physiological mechanisms for aerial respiration in the eel. Comp Biochem Physiol 15:469–484
Bery A, Jensen RA (1988) Biochemical evidence for phylogenetic branching patterns. Bioscience 38(2):103
Bertollo LAC, Takahashi CS, Almeida-Toledo LF, Galetti PM Jr, Ferrari I, Moreira O, Foresti F (1980) Estudos citogenéticos em peixes da região amazônica. I. Ordem Cypriniformes. Ciênc Cult (Suppl)32:735
Bertollo LAC, Takahashi CS, Moreira O (1983) Multiple sex chromosomes in the genus *Hoplias* (Pisces, Erythrinidae). Cytologia 48:1–12
Bicudo JEPW, Johansen K (1979) Respiratory gas exchange in the airbreathing fish, *Synbranchus marmoratus*. Environ Biol Fishes 4:55–64
Bigarella JJ, Ferreira AMM (1985) Amazonian geology and the Pleistocene and the Cenozoic environments and palaeoclimates. In: Prance GT, Lovejoy TE (eds) Amazonia. Key environments. Pergamon Press, Oxford, pp 49–71
Bishop IR, Foxon GEH (1968) The mechanism of breathing in the South American lungfish *Lepidosiren paradoxa*: radiological study. J Zool (Lond) 154:263–272
Bittencourt MM (1991) Exploração dos recursos pesqueiros na Amazônia Central: situação do conhecimento atual. In: Val Al, Feldberg E, Figliuolo R (eds) Bases Científicas para Estratégias de Preservação e Desenvolvimento da Amazônia: Fatos e Perspectivas. Imprensa da Universidade do Amazonas, Manaus, pp 321–326
Böhlke JE, Weitzman SH, Menezes NA (1978) Estado atual da sistemática dos peixes de água doce da América do Sul. Acta Amazonica 8(4):657–677
Bolaffi JL, Booke, HE (1974) Temperature effects on lactate dehydrogenase isozyme distribution in skeletal muscle of *Fundulus heteroclitus* (Pisces: Cyprinodontiformes). Comp Biochem Physiol 48B:557–564
Borges GA (1986) Ecologia de três espécies do gênero *Brycon* Müller & Troschel, 1944 (Pisces-Characidae), no rio Negro-Amazonas, com ênfase na caracterização taxônomica e alimentação. MSc Thesis, PPG INPA/FUA, Manaus, 150 pp
Boutilier RG (1990) Respiratory gas tensions in the environment. In: Boutilier RG (ed) Advances in comparative and environmental physiology, vol 6. Springer, Heidelberg, pp 1–13
Borges GA (1986) Ecologia de três espécies do gênero *Brycon* no rio Negro (Amazonas), com ênfase na caracterização e alimentação. MSc Thesis, PPG INPA/FUA, Manaus, 131 pp
Braum E (1983) Beobachtungen uber eine reversible Lippenextension und ihre Rolle be der Notatmung von *Brycon* spec. (Pisces, Characidae) und *Colossoma macropomum* (Pisces, Serrasalminae). Amazoniana 7:355–374
Braum E, Junk WJ (1982) Morphological adaptation of two Amazonian characoids (Pisces) for surviving in oxygen deficient waters. Int Rev Gesamten Hydrobiol 67:869–886
Brauner CJ, Val AL, Randall DJ (1993) The effect of graded methaemoglobin levels on the swimming performance in chinook salmon (*Oncorhynchus tshawytscha*). J Exp Biol 185: 121–135
Brasil/Eletronorte (1987) UHE Balbina – Fev/87. Centrais Elétricas do Norte do Brasil (ELETRONORTE). Brasília, DF, 26 pp

Britain T (1987) The Root effect. Comp Biochem Physiol 86B:473–481
Brito AL (1981) Aspectos anatômicos e considerações sobre os hábitos de *Pterygoplichthys multiradiatus* Hancock, 1828 do Bolsão do Janauacá-Amazonas, Brasil (Osteichthys, Siluriformes, Loricariidae). MSc Thesis, PPG INPA/FUA, Manaus, 102 pp
Britsky HA (1972) Sistemática e evolução dos Auchenipeteridae e Ageneiosidae (Teleostei, Siluriformes). PhD Thesis, Universidade de São Paulo, 142 pp
Brooks DR, McLennan DA (1991) Phylogeny, ecology, and behavior. Univ Chicago Press, Chicago, 434 pp
Brunori M, Bonaventura J, Focesi A, Galdames-Portus MI, Wilson MT (1979) Separation and characterization of the hemoglobin components of *Pterygoplichthys pardalis*, the Acaribodo. Comp Biochem Physiol 62A:173–178
Buikema AL Jr, Cairns J Jr (1984) Restoration of habitats impacted by oil spills. Butterworth Stoneham, 182 pp
Bunn HF, Riggs A (1979) The measurement of the Bohr effect of fish hemoglobins by gel electrofocusing. Comp Biochem Physiol 62A:95–100
Burggren WW, Johansen K (1987) Circulation and respiration in lungfishes (Dipnoi). J Morphol 1986 (Suppl 1):217–236
Burggren WW, Johansen K, NcMahon B (1985) Respiration in phyletically ancient fishes. In: Forman RE, Gorbman A, Dodd JM, Olsson R (eds) Evolutionary biology of primitive fishes. Plenum Press, New York, pp 217–252
Buschbacher RJ (1987) Deforestation for sovereignty over remote frontiers. Government sponsored pastures in Venezuela near the Brazilian border. In: Jordan CF (ed) Amazonian rain forests. Ecosystem disturbance and recovery. Springer, Berlin Heidelberg New York, pp 46–57
Buschbacher RJ, Uhl C, Serrão EAS (1987) Large-scale development en eastern Amazonia. Pasture management and environmental effects near Paragominas, Pará. In: Jordan CF (ed) Amazonian rain forests. Ecosystem disturbance and recovery. Springer, Berlin Heidelberg New York, pp 90–99
Bushnell P, Jones DR (1992) Cardiovascular and respiratory physiology of tuna. Adaptations for support of exceptionally high metabolic rates. Environ Biol Fishes 40:303–318
Caldas LR (1990) Um pigmento nas águas negras. Ciênc Hoje 11(64):55–57
Cameron JN (1971) Methemoglobin in erythrocytes of rainbow trout. Comp Biochem Physiol 40A:743–749
Cameron JN, Wood CM (1978) Renal function and acid-base regulation in two Amazonian erythrinid fishes: *Hoplias malabaricus*, a water-breather, and *Hoplerythrinus unitaeniatus*, a facultative air-breather. Can J Zool 56:917–930
Caraciolo MCM (1989) Estudos sobre a s-MDH em 12 espécies de peixes da família Curimatidae (Characiformes) da bacia amazônica: aspectos adaptativos e evolutivos. MSc Thesis, PPG INPA/FUA, Manaus, 133 pp
Caraciolo MCM, Val AL, Almeida-Val VMF (1994) Malate dehydrogenase polymorphism in Amazon curimatids (Teleostei: Curimatidae): evidence of an ancient mutational event. Braz J Genet (in press)
Carter GS, Beadle LC (1931a) Notes on the habits and development of *Lepidosiren paradoxa*. J Linn Soc Lond Zool 37:197–203
Carter GS, Beadle LC (1931b) The fauna of the swamps of the Paraguayan Chaco in relation to its environment. I. Physico-chemical nature. J Linn Soc Lond Zool 37:205–258
Carter GS, Beadle LC (1931c) The fauna of the swamps of the Paraguayan Chaco in relation to its environment. II. Respiratory adaptations in the fishes. J Linn Soc Lond Zool 37:327–366
Carvalho FM (1980) Alimentação do mapará (*Hypophthalmus edentatus* Spix, 1829) do lago do Castanho, Amazonas (Siluriformes, Hypophthamidae). Acta Amazonica 10(3):545–555
Carvalho JL, Merona B (1986) Estudo sobre dois peixes migratórios do baixo Tocantins, antes do fechamento da barragem de Tucuruí. Amazoniana 9(4):549–607

Carvalho MF (1981) Alimentação do tambaqui jovem (*Colossoma macropomum* Cuvier, 1818) e sua relação com a comunidade zooplanctonica do lago Grande, Manaquiri, Solimões, AM. MSc Thesis, PPG INPA/FUA, Manaus, 90 pp

Castellini MA, Somero GN (1981) Buffering capacity of vertebrate muscle: correlations with potentials for anaerobic function. J Comp Physiol 143:191–198

Cestari MM, Ferreira R, Galetti PM Jr (1990) Complemento cariotípico de duas espécies de peixes ornamentais: *Chilodus punctatus* (Chilodontidae) e *Anostomus anostomus* (Anostomidae) (Characiformes). III. Simp Citog Evol Aplic Peixes Neotropicais, Abstr Book, p 3

Chaves PTC, Vazzoler AEA de M (1984) Aspectos biológicos de peixes amazônicos. II. Anatomia e microscópica de ovários, escala de maturidade e tipo de desova das espécies do gênero *Semaprochilodus*. Rev Bras Biol 44(3):347–359

Coburn CB, Fischer BA (1973) Red blood cell hematology of fishes: a critique of techniques and a compilation of published data. J Mar Sci 2:37–58

Collins MT, Gratzek JB, Shotts EB, Dane DL, Campbell LM, Senn DR (1975) Nitrification in an aquatic recirculating system. J Fish Res Board Can 32:2025–2031

Coppes Z (1992) Lactate dehydrogenase in teleosts. The role of LDH-C_4 isozyme. Comp Biochem Physiol 102B(4):673–677

Coppes Z, Somero GN (1990) Temperature-adaptive differences between the M_4 lactate dehydrogenases of stenothermal and eurythermal scianid fishes. J Exp Zool 254:127–131

Correa MAV (1987) Crescimento do matrinchã, *Brycon cephalus* (Gunther, 1869) (Teleostei, Characidae) no baixo rio Negro, seus afluentes e no baixo rio Solimões. MSc Thesis, PPG INPA/FUA, Manaus, 124 pp

Costa OTF (1991) Parâmetros hemoglobínicos de *Hoplosternum littorale* sob exposição ao petróleo do rio Urucú. relatório preliminar CNPq. Internal Rep, 48 pp

Cox-Fernandes C (1989) Estudo de migrações laterais de peixes do Sistema lago do Rei (Ilha do Careiro), AM. Brasil. MSc Thesis, PPG INPA/FUA, Manaus, 170 pp

Cox-Fernandes C, Petry P (1991) A importância da várzea no ciclo de vida dos peixes migradores na Amazônia central. In: Val AL, Feldberg E, Figliuolo R (eds) Bases Científicas para Estratégias de Preservação e Desenvolvimento da Amazônia: Fatos e Perspectivas. Imprensa da Universidade do Amazonas, Manaus, pp 315–320

Cuffney TF (1988) Input, movement and exchange of organic matter within a subtropical coastal blackwater river-floodplain system. Freshwater Biol 19:305–320

Dafré AL, Wilhelm D (1989) Root effect hemoglobins in marine fish. Comp Biochem Physiol 92A:467–471

D'Ávila-Limeira NC (1989) Estudos sobre a lactato desidrogenase (LDH) em 27 espécies de peixes da bacia Amazônica: aspectos adaptativos e evolutivos. MSc Thesis, PPG INPA/FUA, Manaus, 123 pp

Davis JC (1975) Minimal dissolved oxygen requirements of aquatic life with emphasis on Canadian species: a review. J Fish Res Board Can 32(12):2295–2332

De Haan H, Jones RI, Solonen K (1990) Abiotic transformation of iron and phosphate in humic lake water by double-isotope labelling and gel filtration. Limnol Oceanogr 35(2): 491–497

Dejours P (1979) Oxygen demand and gas exchange. In: Wood SC, Lenfant C (eds) Evolution of respiratory processes. Marcel Dekker, New York, pp 1–49

DeLaney RG, Laurent P, Galante R, Pack AI, Fishman AP (1983) Pulmonary mechanoreceptors in the dipnoi lungfish *Protopterus* and *Lepidosiren*. Am J Physiol 244:R418–R428

Denison RH (1941) The soft anatomy of *Bothriolepis*. J Paleontol 15:553–561

Dourojeanni MJ (1985) Over-exploited and under-used animals in the Amazon region. In: Prance GT, Lovejoy TE (eds) Amazonia. Key environments. Pergamon Press, Oxford, pp 419–433

Driedzic WR, Phleger CF, Fields JHA, French C (1978) Alterations in energy metabolism associated with the transitions from water to air breathing in fish. Can J Zool 56:730–735

Dunn JF, Hochachka PW, Davison W, Guppy M (1983) Metabolic adjustments to diving and recovery in the African lungfish. Am J Physiol 245:R651–R657

Duran N (1990) Violaceína: a descoberta de um antibiótico. Ciênc Hoje 11(64):58–60

Else PL, Hulbert AJ (1985) An allometric comparison of the mitochondria of mammalian and reptilian tissues: the implications for the evolution of endothermy. J Comp Physiol 156B: 3–11

Ertel JR, Hedges JI, Devol AH, Richey JE (1986) Dissolved humic substances of the Amazon river system. Limnol Oceanogr 31(4):739–754

Esteves FA, Bozelli RL, Roland F (1990) Lago Batata: um laboratório de limnologia tropical. Ciênc Hoje 11(64):26–33

Everse J, Kaplan NO (1973) Lactate dehydrogenase: structure and function. Adv Enzymol Relat Areas Mol Biol 37:61–148

Falesi IC (1986) Estado atual de conhecimentos sobre os solos da Amazônia brasileira. Anais do 1° Simpósio sôbre o Trópico Úmido, pp 168–191

Farber J, Rahn H (1970) Gas exchange between air and water and the ventilation pattern in the electric eel. Respir Physiol 9:151–161

Farias IP (1992) Efeito da aclimatação térmica sobre a lactato desidrogenase de *Colossoma macropomum* e *Hoplosternum littorale* (Amazonas, Brasil). MSc Thesis, PPG INPA/FUA, Manaus, 72 pp

Farias IP, Almeida-Val VMF (1992) Malate dehydrogenase (sMDH) in Amazon cichlid fishes: evolutionary features. Comp Biochem Physiol 103B:939–943

Farmer M, Fyhn HJ, Fyhn UEH, Noble RW (1979) Occurrence of Root effect hemoglobins in Amazonian fishes. Comp Biochem Physiol 62A:115–124

Farrell AP (1978) Cardiovascular events associated with air-breathing in two teleosts, *Hoplerythrinus unitaeniatus* and *Arapaima gigas*. Can J Zool 56:953–958

Farrell AP, Randall DJ (1978) Air-breathing mechanics in two Amazonian teleosts, *Arapaima gigas* and *Hoplerythrinus unitaeniatus*. Can J Zool 56:939–945

Fearnside PM (1989) Deforestation and international economic development projects in Brazilian Amazonia. In: Schneider A (ed) Deforestation and "development" in Canada and the Tropics. The impact on people and environment. University of Cape Breton, Sydney, pp 23–27

Fearnside PM (1990) A hidrelétrica de Balbina: o faraonismo irreversível versus o meio ambiente na Amazônia. Instituto de Antropologia e Meio Ambiente, São Paulo, 69 pp

Feldberg E (1990) Estudos citogenéticos em 12 espécies de peixes da família Curimatidae (Characiformes) da Amazônia central. PhD Thesis, PPG INPA/FUA, Manaus, 126 pp

Feldberg E, Bertollo LAC (1985) Nucleolar organizing regions in some species of neotropical cichlids fish (Pisces, Perciformes). Caryologia 38(3/4):319–324

Feldberg E, Bertollo LAC, Almeida-Toledo LF, Foresti F, Moreira O, Santos AF (1987) Biological aspects of amazonian fishes. IX. Cytogenetic studies in two species of the genus *Semaprochilodus* (Pisces, Prochilodontidae). Genome 29(1):1–4

Feldberg E, Porto JIR, Bertollo LAC (1992) Karyotype evolution in Curimatidae (Teleostei, Characiformes) of the Amazon region. I. Studies on the genera *Curimata*, *Psectrogaster*, *Steindachnerina* and *Curimatella*. Brazil J Genet 15(2):369–383

Fennocchio AS, Bertollo LAC (1987) Estudos citogenéticos preliminares de Siluriformes da bacia Amazônica. Ciênc Cult 39(Suppl):732

Ferguson MM, Allendorf FW (1991) Evolution of fish genome. In: Hochachka PW, Mommsen TP (eds) Biochemistry and molecular biology of fishes, vol 1. Elsevier, Amsterdam, pp 25–42

Ferguson RA, Boutilier RG (1989) Metabolic-membrane coupling in red blood cells of trout: the effects of anoxia and adrenergic stimulation. J Exp Biol 143:149–164

Fernandes CC (1988) Estudos de migrações laterais de peixes no sistema lago do Rei (Ilha do Careiro)-AM, BR. MSc Thesis, PPG INPA/FUA, Manaus, 170 pp

Ferreira EJG (1981) Alimentação dos adultos de doze espécies de cichlídeos (Perciformes, Cichlidae) do rio Negro, Brasil. MSc Thesis, PPG INPA/FUA, Manaus, 254 pp

Ferreira EJG (1986) Estudos e levantamento do impacto ambiental da UHE de Cachoeira Porteira. Relatório do Subprojeto "Identificção e descrição das principais espécies de peixes existentes". Convênio Engerio/CNPq-INPA, Manaus, 99 pp
Ferreira EJG, Santos GM, Jegu M (1988) Aspectos ecológicos da ictiofauna do rio Mucajaí, na área da ilha Paredão, Roraima, Brasil. Amazoniana 10(3):339–352
Ferreira RCH, Apel LE (1990) Estudo detalhado de fontes e usos de mercúrio. Relatório preliminar. Cent Tecnol Miner, Rio de Janeiro, 62 pp
Ferris SD, Whitt GS (1979) Evolution of the differential regulation of duplicate genes after polyploidization. J Mol Evol 12:267–317
Fink SV, Fink WL (1981) Interrelationships of the ostariophysan fishes (Teleostei). Zool J Linn Soc 72:297–353
Fink WL, Fink SV (1979) Central Amazonia and its fishes. Comp Biochem Physiol 62A:13–29
Fishman AP, Galante RJ, Pack AI (1989) Diving physiology. Lungfish. In: Wood SC (ed) Comparative pulmonary physiology. Marcel Dekker, New York, pp 645–676
Fittkau EJ (1975) Ökologische Gliederung des Amazonasgebietes auf geochemischer Grundlage. Munster Forsch Geol Paläontoe 20/21:35–50
Fittkau EJ (1985) Ökologische und faunenhistorische Zoogeographie der tropischen Regenwälder: Versuch eines Vergleichs (Ecological and historical zoogeography of tropical rainforests: attempt of a comparison). Verh Dtsch Zool Ges 78:137–146
Fontenele O (1953) Contribuição para o conhecimento da biologia do apaiari *Astronotus ocellatus* (Spix) (Cichlidae) em cativeiro. Rev Bras Biol 11:467–484
Foresti F, Carneito MILC, Nakamoto W (1982) Estudos cromossômicos em populações de *Synbranchus marmoratus* Bloch, 1795 (Pisces, Synbranchidae). Ciênc Cult 34(Suppl 7): 747
Fowler HW (1951) Os peixes de água doce do Brasil. Departamento de Zoologia da Secretaria de Agricultura do Estado de São Paulo, São Paulo, 625 pp
Foxon GEH (1950) A description of the coronary arteries in dipnoan fishes and some remarks on their importance from the evolutionary standpoint. J Anat 84:121–131
Frailey CD, Lavina EL, Rancy A, Souza JP (1988) A proposed pleistocene/holocene lake in the Amazon basin and its significance Amazonian geology and biogeography. Acta Amazonica 18(3):119–143
French CJ, Hochachka PW (1978) Lactate dehydrogenase isozymes from heart and white muscle of water-breathing and air-breathing fish from the Amazon. Can J Zool 56:769–773
Fry FE, Hochachka PW (1970) Fish. In: Whitton GC (ed) Comparative physiology of thermoregulation, vol 1. Academic Press, New York, pp 79–134
Fullarton MH (1931) Notes on the respiration of *Lepidosiren*. Proc Zool Soc Lond 1950: 1301–1306
Furch K (1984) Water chemistry of the Amazon basin: the distribution of chemical elements among fresh waters. In: Sioli H (ed) The Amazon. Limnology and landscape ecology of a mighty tropical river and its basin. Junk, Dordrecht, pp 167–200
Furch K, Junk WJ, Klinge H (1982) Unusual chemistry of natural waters from the Amazon region. Acta Cient Venez 33:261–273
Fyhn UEH, Fyhn HJ, Davis BJ, Powers DA, Fink WL, Garlick RL (1979) Hemoglobin heterogeneity in Amazonian fishes. Comp Biochem Physiol 62A:39–66
Galdames-Portus MI, Noble RW, Farmer M, Powers DA, Riggs A, Brunori M, Fyhn HJ, Fyhn UEH (1979) Studies of the functional properties of the hemoglobins of *Ostegolossum bicirrhosum* and *Arapaima gigas*. Comp Biochem Physiol 62A:145–154
Galdames-Portus MI, Donald EL, Focesi A Jr (1982) Hemoglobinas em silurídeos da Amazônia Central. I. Análise eletroforética dos hemolisados. Acta Amazônica 12(4):701–711
Galetti PM Jr, Mestriner CA, Venere PC, Foresti F (1991) Heterochromatin and karyotype reorganization in fish of the family Anostomidae (Characiformes). Cytogen Cell Genet 56:116–121
Galvis G, Mojica JI, Rodrigues F (1989) Estudio ecologico de una laguna de desborde del rio Metica, Oronoquia Colombiana. Universidad Nacional del Colombia, 164 pp

Gans C (1970) Strategy and sequence in the evolution of the external gas exchangers of ectothermal vertebrates. Forma Functio 3:61–104
Garavello JC (1979) Revisão taxonômica do gênero *Leporinus*. PhD Thesis, USP, São Paulo, 450 pp
Garey WF, Rahn H (1970) Normal arterial gas tensions and pH and the breathing frequency of the electric eel. Resp Physiol 9:141–150
Garlick RL, Bunn HF, Fyhn HJ, Fyhn UEH, Martin JP, Noble RW, Powers DA (1979) Functional studies on the separated hemoglobin components of an air-breathing catfish, *Hoplosternum littorale* (Hancock). Comp Biochem Physiol 62A:219–226
Gee JH (1976) Buoyancy and aerial respiration: factors influencing the evolution of reduced swim-bladder volume of some Central American Catfishes (Trichomycteridae, Callichthyidae, Loricariidae, Astroblepidae). Can J Zool 54:1030–1037
Gentilli J (1949) Foundation of Australian bird geography. Emu 49:85–130
Géry J (1969) The fresh-water fishes of South America. In: Fittkau EJ, Illies J, Klinge H, Schwabe GH, Sioli H (eds) Biogeography and ecology in South America. Junk, The Hague, pp 828–848
Géry J (1984) The fishes of Amazonia. In: Sioli H (ed) The Amazon. Limnology and landscape ecology of a mighty tropical river and its basin. Junk, Dordrecht, pp 15–46
Gesser RH, Poupa O (1973) The lactate dehydrogenase system in the heart and skeletal muscle of fish: a comparative study. Comp Biochem Physiol 46B:683–690
Gillen RG, Riggs A (1971) The hemoglobins of a fresh-water teleost, *Cichlasoma cyanoguttatun* (Baird & Girard). I. The effects of phosphorylated organic compounds upon the oxygen equilibria. Comp Biochem Physiol 38B:585–595
Gillen RG, Riggs A (1977) The enhancement of the alkaline Bohr effect of some fish hemoglobins with adenosine triphosphate. Arch Biochem Biophys 183:678–685
Giuliano-Caetano L, Bertollo LAC (1988) Karyotype variability in *Hoplerythrinus unitaeniàtus* (Characiformes, Erythrinidae). I. Chromosome polymorphism in the rio Negro population (Manaus, state of Amazonas). Brazil J Genet 11(2):299–306
Giuliano-Caetano L, Bertollo LAC (1990) Karyotype variability in *Hoplerythrinus unitaeniatus* (Pisces, Characiformes, Erythrinidae). II. Occurrence of natural triploidy. Brazil J Genet 13(2):231–237
Glass ML (1992) Ventilatory responses to hypoxia in ectothermic vertebrates. In: Wood SC, Weber RE, Hargens AR, Millard RW (eds) Physiological adaptations in vertebrates: respiration, circulation, and metabolism. Marcel Dekker, New York, pp 97–118
Godoy MP (1979) Marcação e migração de piramutaba *Brachyplatystoma vaillanti* (Val, 1840) na bacia Amazônica (Pará e Amazonas), Brasil (Pisces, Mematognathi, Pimelodidae). Bol FCAP Belém 11:3–21
Goldberg E, Wuntch T (1967) Electrophoretic and kinetic properties of *Rana pipiens* LDH isozymes. J Exp Zool 165:101
Goulding M (1979) Ecologia da Pesca no rio Madeira. CNPq/INPA, Manaus, 172 pp
Goulding M (1980) The fishes and the forest. Explorations in Amazonian natural history. Univ California Press, Berkeley, 280 pp
Goulding M (1985) Forest fish of the Amazon. In: Prance GT, Lovejoy TE (eds) Amazonia. Key environments. Pergamon Press, Oxford, pp 267–276
Goulding M (1989) Amazon. The flooded forest. BBC Books, London, 208 pp
Goulding M, Carvalho ML, Ferreira EJG (1988) Rio Negro-rich life in poor water: Amazonia diversity and foodchain ecology as seen through fish communities. SBP Publ, The Hague, 200 pp
Gradwell N (1971a) A photographic analysis of the air breathing behavior of the catfish, *Plecostomus punctatus*. Can J Zool 49:1089–1094
Gradwell N (1971b) A muscular oral valve unique in fishes. Can J Zool 49:837–839
Graves JE, Somero GN (1982) Electrophoretic and functional enzymic evolution in four species of eastern Pacific barracudas from different thermal environments. Evolution 36:97–106

Greaney DS, Powers DA (1978) Allosteric modifiers of fish hemoglobins: in vitro and in vivo studies of the effect of ambient oxygen and pH on erythrocytic ATP concentrations. J Exp Zool 203:339–350
Greenwood PH, Rosen DE, Weitzman SH, Mayers GS (1966) Phyletic studies of teleostean fishes, with a provisional classification of living forms. Bull Am Nus Nat Hist 131:339–456
Gronenborn AM, Clore GM, Brunori M, Giardina B, Falcioni G, Perutz MF (1984) Stereochemistry of ATP and GTP bound to fish hemoglobins. A transferred nuclear Overhauser enhancement. ^{31}P-nuclear magnetic resonance, oxygen equilibrium and molecular modeling study. J Mol Biol 178:731–742
Gutiérrez M, Sarrasquete MC (1983) Dados sobre eritrócitos de varias especies de peces teleósteos marinos de la costa subatlántica de la España. Invest Pesq 47:85–94
Hara TJ, Law YMC, MacDonald S (1976) Effects of mercury and copper on the olfactory response in rainbow trout, *Salmo gairdneri*. J Fish Res Board Can 33:1568–1573
Hardy ER (1980) Composição do zooplancton em 5 lagos da Amazônia Central. Acta Amazonica 10:577–609
Hauss R (1975) Wirkungen der Temperaturen auf Proteine. Enzyme und Isozenzyme aus Organen des Fisches *Rhodeus amarus*. Bloch. I. Einfluß der Adaptationstemperatur auf die weiße Rückenmuskulatur. Zool Anz 194:243–261
Hazel J, Prosser CL (1974) Molecular mechanisms of temperature compensation in poikilotherms. Physiol Rev 54:620–677
Herbert CV, Jackson DC (1985) Temperature effects on the responses to prolonged submergence in the turtle *Chrysemys picta bellii*. II. Metabolic rate, blood acid-base and ionic changes, and cardiovascular function in aerated and anoxic water. Physiol Zool 58:670–681
Heisler N (1982) Intracellular and extracellular acid-base regulation in the tropical fresh-water teleost fish *Synbranchus marmoratus* in response to the transition from water breathing to air breathing. J Exp Biol 99:9–28
Hinegardner R (1976) Evolution of genome size. In: Ayala FJ (ed) Molecular evolution. Sinauer, Sunderland, pp 179–199
Hinergardner R, Rosen DE (1972) Cellular DNA content and the evolution of Teleostean fishes. Am Nat 106:621–644
Hochachka PW (1965) Isoenzymes in metabolic adaptation of poikilotherms: subunit relationships in lactic dehydrogenases of goldfish. Arch Biochem Biophys 111:96–103
Hochachka PW (1979) Cell Metabolism, air breathing, and the origins of endothermy. In: Wood SC, Lenfant C (eds) Evolution of respiratory processes. Marcel Dekker, New York, pp 253–288
Hochachka PW (1980) Living without oxygen. Harvard Univ Press, Cambridge, 181 pp
Hochachka PW (1988a) The nature of evolution and adaptation: resolving the unity-diversity paradox. Can J Zool 66:1146–1152
Hochachka PW (1988b) Metabolic key to hypoxia tolerance: loss of regulatory links between the electron transfer system and glycolysis. In: Lemasters JJ, Hackenbrock CR, Thurman RG, Westerhoff HV (eds) Integration of mitochondrial function. Plenum, New York, pp 623–627
Hochachka PW (1993) Adaptability of metabolic efficiencies under chronic hypoxia in man. In: Hochachka PW, Lutz PL, Sick T, Rosenthal M, Thillart van den G (eds) Surviving hypoxia: mechanisms of control and adaptation. CRC Press, Boca Raton, pp 127–135
Hochachka PW, Clayton-Hochachka B (1973) Glucose-6-phosphate dehydrogenase and thermal acclimation in the mullet fish. Mar Biol 18:251–259
Hochachka PW, Guppy M (1987) Metabolic arrest and the control of biological time. Harvard Univ Press, Cambridge, 227 pp
Hochachka PW, Hulbert WC (1978) Glycogen "seas", glycogen bodies, and glycogen granules in heart and skeletal muscle of two air-breathing, burrowing fishes. Can J Zool 56:774–786
Hochachka PW, Lewis JK (1971) Interacting effects of pH and temperature on the Km values for fish tissue lactate dehydrogenases. Comp Biochem Physiol 39B:925–933

Hochachka PW, Randall DJ (1978) Alpha-Helix Amazon expedition, September–October 1976. Can J Zool 56:713–716
Hochachka PW, Somero GN (1968) The adaptation of enzymes to temperature. Comp Biochem Physiol 27:659–668
Hochachka PW, Somero GN (1973) Strategies of biochemical adaptation. Saunders, Philadelphia, 385 pp
Hochachka PW, Somero GN (1984) Biochemical adaptation. Princeton Univ Press, Princeton, 537 pp
Hochachka PW, Moon TW, Bailey J, Hulbert WC (1978a) The osteoglossid kidney: correlations of structure, function, and metabolism with transition to air-breathing. Can J Zool 56:820–832
Hochachka PW, Guppy M, Guderley KB, Storey KB, Hulbert WC (1978b) Metabolic biochemistry of water- vs. air-breathing fishes: muscle enzymes and ultrastructure. Can J Zool 56:736–750
Hochachka PW, Guppy M, Guderley KB, Storey KB, Hulbert WC (1978c) Metabolic biochemistry of water- vs. air-breathing osteoglossids: heart enzymes and ultrastructure. Can J Zool 56:759–768
Hochachka PW, Hulbert WC, Guppy M (1979) The tuna power plant and furnace. In: Sharp GD, Dizon AE (eds) The physiological ecology of tunas. Academic Press, New York, pp 153–181
Hochachka PW, Matheson GO, Parkhouse WS, Sumar-Kalinowski J, Stanley C, Monge C, McKenzie DC, Merkt J, Man SFP, Jones R, Allen PS (1991) Inborn resistence to hypoxia in high altitude. In: Lahiri S, Cherniak NS, Fitzgerald RS (eds) Responses and adaptation to hypoxia: organ to organelle. Oxford Univ Press, New York, pp 191–194
Holeton GF, Stevens ED (1978) Swimming energies of an Amazonian characin in "black" and "white" water. Can J Zool 56:983–987
Hollanda OM (1982) Captura, distribuição, alimentação e aspectos reprodutivos de *Hemiodus unimaculatus* (Bloch, 1794) e *Hemiodopsis* sp. (Osteichthyes, Characoidei, Hemiodidae) na represa hidrelétrica de Curua-Una, Pará. MSc Thesis, PPG INPA/FUA, Manaus, 99 pp
Honda EMS, Correa CM, Castelo FP, Zapelini EA (1975) Aspectos gerais do pescado no Estado do Amazonas. Acta Amazonica 5:87–94
Houston AH (1985) Erythrocytic magnesium in freshwater fishes. Magnesium 4:106–128
Houston AH, Koss TF (1984) Erythrocytic haemoglobin, magnesium and nucleoside triphosphate levels in rainbow trout exposed to progressive heat stress. J Therm Biol 9:159–164
Hughes GM, Singh BN (1971) Gas exchange with air and water in an air-breathing catfish *Saccobranchus fossilis*. J Exp Biol 55:667–682
Hulbert WC, Moon TW, Hochachka PW (1978a) The osteoglossid gill: correlations of structure, function, and metabolism with transition to air-breathing. Can J Zool 56:801–808
Hulbert WC, Moon TW, Hochachka PW (1978b) The erythrinid gill: correlations of structure, function, and metabolism. Can J Zool 56:814–819
Ingermann RL, Terwilleger RC (1981) Oxygen affinities of fetal and maternal hemoglobins of the viviparous seaperch *Embiotoca lateralis*. J Comp Physiol 142:523–531
Ingermann RL, Terwilleger RC (1982) Presence and possible function of Root effect hemoglobins in fishes lacking functional swimbladder. J Exp Zool 220:171–177
INPA/ENGE-RIO (1986) Estudos e levantamentos do impacto ambiental da UHE de Cachoeira Porteira. Sub-projeto: identifiação e descrição das principais espécies de peixes existentes. Relatório Técnico, INPA, Manaus, 60 pp
Isaacks RE, Harkness DR (1980) Erythrocytic organic phosphates and hemoglobin function in birds, reptiles, and fishes. Am Zool 20:115–129
Isaacks RE, Kim HD, Bartlett GR, Harkness DR (1977) Inositol pentaphosphate in erythrocytes of a fresh water fish, pirarucú (*Arapaima gigas*). Life Sci 20:987–990

Isaacks RE, Kim HD, Harkness DR (1978) Inositol diphosphate in erythrocytes of the lungfish, *Lepidosiren paradoxa*, and 2,3 diphosphoglycerate in erythrocytes of the armored catfish, *Pterygoplichthys* sp. Can J Zool 56:1014–1016

Isaia J, Maetz J, Haywood GP (1978) Effects of epinephrine on branchial nonelectrolyte permeability in trout. J Exp Biol 46:317–327

Iwama GK, Boutilier RG, Heming TA, Randall DJ, Mazeaud M (1987) The effects of altering gill water flow on gas transfer in rainbow trout. Can J Zool 65:2466–2470

Jenkins RE (1976) A list of undescribed freshwater fish of continental United States and Canada, with additions to the 1970 checklist. Copeia 1976(3):642–644

Johansen K (1966) Air breathing in the teleost *Synbranchus marmoratus*. Comp Biochem Physiol 18:383–395

Johansen K (1970) Air breathing in fishes. In: Hoar WS, Randall DJ (eds) Fish physiology. Academic Press, New York, pp 361–411

Johansen K (1979) Cardiovascular support of metabolic functions in vertebrates. In: Wood SC, Lenfant C (eds) Evolution of respiratory processes: a comparative approach. Marcel Dekker, New York, pp 107–192

Johansen K (1982) Respiratory gas exchange of vertebrate gills. In: Houlihan DF, Rankin JC, Shuttleworth TJ (eds) Gills. Cambridge Univ Press, Cambridge, pp 99–128

Johansen K, Lenfant C (1967) Respiratory function in the South American lungfish, *Lepidosiren paradoxa*. J Exp Biol 46:205–218

Johansen K, Lenfant C (1968) Respiration in the African lungfish *Protopterus aethiopicus*. II. Control of breathing. J Exp Biol 49:453–468

Johansen K, Lenfant C (1972) A comparative approach to the adaptability of O_2-Hb affinity. In: Astrup P, Rorth M (eds) Oxygen affinity of hemoglobin and red cell acid-base status. Academic Press, New York, pp 750–780

Johansen K, Lenfant C, Grigg GC (1967) Respiratory control in the lungfish, *Neoceratodus forsteri* (Krefft). Comp Biochem Physiol 20:835–854

Johansen K, Lenfant C, Schmidt-Nielsen K, Petersen JA (1968) Gas exchange and control of breathing in the electric eel, *Electrophorus electricus*. Z Vgl Physiol 61:137–163

Johansen K, Hansen D, Lenfant C (1970) Respiration in a primitive air breather, *Amia calva*. Respir Physiol 9:162–174

Johansen K, Mangum CP, Lykkeboe G (1978a) Respiratory properties of the blood of Amazon fishes. Can J Zool 56:898–906

Johansen K, Mangum CP, Weber RE (1978b) Reduced blood O_2 affinity associated with air-breathing in osteoglossid fishes. Can J Zool 56:891–897

Johnsen PB (1984) Establishing the physiological and behavioral determinants of chemosensory orientation. In: McCleave JD, Arnold GP, Dodson JJ, Neil WH (eds) Mechanisms of migration in fishes. Proc of a NATO advanced research on mechanisms of migration in fishes. Plenum Press, New York, pp 379–385

Johnson LF, Tate ME (1969) Structure of "phytic acids". Can J Chem 47:63–73

Jones FRH (1968) Fish migration. Edward Arnold, London, 325 pp

Jones JRE (1964) Fish and river pollution. Butterworth, London, 203 pp

Jordan CF (1987) Shifting cultivation. Slash and burn agriculture near San Carlos de Rio Negro. In: Jordan CF (ed) Amazonian rain forests. Ecosystem disturbance and recovery. Springer, Berlin Heidelberg New York, pp 9–23

Junk WJ (1970) Investigations on the ecology and production-biology of the "floating-meadows" (*Paspalo-Echinocloetum*) on the middle Amazon. Part I: The floating vegetation and its ecology. Amazoniana 2:449–495

Junk WJ (1973) Investigations on the ecology and production-biology of the "floating-meadows" (Paspalum-Echinocloetum) on the middle Amazon. II. The aquatic fauna in the root-zone of floating vegetation. Amazoniana 4:9–102

Junk WJ (1979) Macrófitas aquáticas nas várzeas da Amazônia e possibilidades de seu uso na agropecuária. Manaus, AM, INPA, 24 pp

Junk WJ (1980) Áreas inundáveis – um desafio para a limnologia. Acta Amazonica 10(4): 775–796
Junk WJ (1983) As águas da região amazônica. In: Salati E, Junk WJ, Schubart HOR, Oliveira de AE (eds) Amazônia: desenvolvimento, integração e ecologia. Editôra Brasiliense and CNPq, pp 45–100
Junk WJ (1984a) Ecology of the várzea, floodplain of Amazonian whitewater rivers. In: Sioli H (ed) The Amazon. Limnology and landscape ecology of a mighty tropical river and its basin. Junk, Dordrecht, pp 215–244
Junk WJ (1984b) Ecology, fisheries and fish culture in Amazonia. In: Sioli H (ed) The Amazon. Limnology and landscape ecology of a mighty tropical river and its basin. Junk, Dordrecht, pp 443–476
Junk WJ, Furch K (1985) The physical and chemical properties of Amazonian waters and their relationships with the biota. In: Prance GT, Lovejoy TE (eds) Amazonia. Key environments. Pergamon Press, Oxford, pp 3–17
Junk WJ, Honda EMS (1976) A Pesca na Amazônia. Aspectos ecológicos e econômicos. Anais I Encontro Nacional de Limnologia, Piscicultura e Pesca Continental, pp 221–226
Junk WJ, Howard-Williams C (1984) Ecology of aquatic macrophytes in Amazonia. In: Sioli H (ed) The Amazon. Limnology and landscape ecology of a mighty tropical river and its basin. Junk, Dordrecht, pp 269–294
Junk WJ, Soares GM, Carvalho FM (1983) Distribution of fish species in a lake of the Amazon river floodplain near Manaus (Lago Camaleão), with special reference to extreme oxygen conditions. Amazoniana 7(4):397–431
Junk WJ, Bailey PB, Sparks RE (1989) The flood-pulse concept in river-floodplain systems. In: Dodge D (ed) Proc Int Large River Symp. Can Spec Publ Fish Aquat Sci 106:110–127
Keenleyside MHA (1979) Diversity and adaptation in fish behaviour. Springer, Berlin Heidelberg New York, 209 pp
Kettler MK, Whitt GS (1986) An apparent progressive and recurrent evolutionary restricution in tissue expression of a gene, the lactate dehydrogenase-C gene within a family of bony fish (Salmoniformes: Umbridae). J Mol Evol 23:95–107
King MC, Wilson AC (1975) Evolution at two levels in humans and chimpanzees. Science 188:107–116
Kirpichnikov VS (1981) The evolution of kariotypes among Cyclostomes and fishes. In: Genetic bases of fish selection. Springer, Berlin New York, 410 pp
Kirpichnikov VS (1992) Adaptive nature of intrapopulational biochemical polymorphism in fish. J Fish Biol 40:1–16
Klammer G (1984) The relief of the extra-Andean Amazon basin. In: Sioli H (ed) The Amazon. Limnology and landscape ecology of a mighty tropical river and its basin. Junk, Dordrecht, pp 47–83
Knöppel HA (1970) Food of Central Amazonian fishes. Amazoniana 2(3):257–352
Koshland DE (1973) Protein shape and biological control. Sci Am 229:52–64
Kramer B (1990) Electro-communication in teleost fishes. Springer, Berlin Heidelberg New Yrok, 240 pp
Kramer DL (1978) Ventilation of the respiratory gas bladder in *Hoplerythrinus unitaeniatus* (Pisces, Characoidei, Erythrinidae). Can J Zool 56:931–938
Kramer DL (1988) The behavioral ecology of air-breathing by aquatic animals. Can J Zool 66:89–94
Kramer DL, Graham JB (1976) Synchronous air breathing, a social component of respiration in fishes. Copeia 1976:689–697
Kramer DL, McClure M (1980) Aerial respiration in the catfish, *Corydoras aeneus* (Callichthyidae). Can J Zool 58:1984–1991
Kramer DL, McClure M (1981) The transit cost of aerial respiration in the catfish *Corydoras aeneus* (Callichthyidae). Physiol Zool 54:189–194

Kramer DL, Lindsey CC, Moodie GEE, Stevens ED (1978) The fishes and the aquatic environment of the central Amazon basin, with particular reference to respiratory patterns. Can J Zool 56:717–729

Krogh A, Leitch I (1919) The respiratory function of the blood in fishes. J Physiol (Lond) 52:288

Kuhn W, Ramel A, Kuhn HJ, Marti E (1963) The filling mechanism of the swimbladder. Generation of high gas pressures through hairpin countercurrent multiplication. Experientia 19:497–552

Kullander SO (1986) Cichlid fish of the Amazon river drainage of Peru. Swedish Museum of Natural History, Stockholm, Sweden, 431 pp

Kullander SO, Ferreira EJG (1988) A new *Satanoperca* species (Teleostei, Cichlidae) from the Amazon river basin in Brazil. Cybium 12(4):343–355

Lacerda LD, Salomons W (1991) Mercury in the Amazon: a chemical time bomb? Report sponsored by Dutch Ministry of Housing, Physical Planning and Environment, 46 pp

Lauder GV, Liem KF (1983a) Patterns of diversity and evolution in ray-finned fishes. In: Northcut RG, Davis RE (eds) Fish neurobiology. Univ Michigan Press, Ann Arbor, pp 1–24

Lauder GV, Liem KF (1983b) The evolution and interrelationships of the Actinopterygian fishes. Bull Mus Comp Zool 150(3):95–197

Lauzanne L, Loubens G (1985) Peces del rio Mamoré. ORSTON, ORSTON/CORDEBENI/ UTB, Paris, 116 pp

Leibel WS, Peairs KS (1990) Molecular homology of coexpressed eye- and liver-specific lactate dehydrogenase isozymes from the basketmouth Cichlid assessed by oxamate-sepharose and blue-dextran-sepharose affinity chromatography. J Exp Zool 253:107–114

Leibel WS, Bass KN, Eshbach JE (1991) An immunochemical analysis of the novel liver-restricted LDH activity from cichlid fishes (Cichlidae: Teleostei) confirms its non-orthology with piscine LDH-C. J Fish Biol 39:143–150

Leigh EG, Wright SJ (1990) Barro Colorado island and tropical biology. In: Gentry AH (ed) Four tropical rainforests. Yale Univ Press, New Haven, pp 28–47

Leite RAN, Bittencourt MM (1991) Impacto das hidrelétricas sobre a ictiofauna da Amazônia: o exemplo de Tucuruí. In: Val AL, Feldberg E, Figliuolo R (eds) Bases científicas para estratégias de preservação e desenvolvimento da Amazônia: fatos e perspectivas. Imprensa da Universidade do Amazonas, Manaus pp, 85–100

Leite RG (1987) Alimentação e hábitos alimentares dos peixes do rio Uatumã na área de abrangência da usina hidrelétrica Balbina, Amazonas, Brasil. MSc Thesis, PPG INPA/FUA, Manaus, 81 pp

Lenfant C, Johansen K, Hansen D (1970) Bimodal gas exchange and ventilation-perfusion relationships in lower vertebrates. Fed Proc 29:1124–1129

Leopoldo PR, Franken W, Matsui E, Salati E (1982) Estimativa da evapotranspiração da floresta Amazônica de terra firme. Acta Amazonica 12(3):23–28

Li WH (1983) Evolution of duplicate genes and pseudogenes. In: Nei M, Koehn RK (eds) Evolution of genes and proteins. Sinauer, Suderland, pp 14–37

Liem KF (1987) Functional design of the air ventilation apparatus and overland excursions by teleost. Fieldiana 37:1–29

Lin H, Randall DJ (1990) The effect of varying water pH on the acidification of expired water in ranbow trout. J Exp Biol 149:149–160

Lock RAC, Cruijsen PM, Van Overbeeke AP (1981) Effects of mercuric chloride and methylmercuric chloride on the osmoregulatory function of the gills in rainbow trout, *Salmo gairdneri*. Comp Biochem Physiol 68C:151–159

Loubens G, Aquim JL (1986) Sexualidad y reproduccion de los principales peces de la cuenca del rio Mamore, Beni, Bolivia. Convenio ORSTOM/UTB/CORDEBENI, Informe cientifico No 5

Love RM (1980) The chemical biology of fishes, vol 2. Academic Press, London, 943 pp

Lowe-McConnell RH (1964) The fishes of the Rupununi savana district of British Guiana. PtI. Groupings of fish species and effects of the seasonal cycles on the fish. Biol J Linn Soc Zool 45:103–144

Lowe-McConnell RH (1975) Fish communities in tropical freshwaters. Longman, New York, 284 pp

Lowe-McConnell RH (1987) Ecological studies in tropical fish communities. Cambridge Univ Press, Cambridge, 382 pp

Lüling KH (1964) Zur Biologie und Ökologie von *Arapaima gigas* (Pisces: Osteoglossidae). Z Morph Ökol Tiere 54:436–530

Lundberg JG, Mago-Leccia F (1986) A review of *Rhabdolichops* (Gymnotiformes, Sternopygidae), a genus of South American freshwater fishes, with descriptions of four new species. Proc Natl Acad Sci Philadelphia 138(1):53–85

Lundberg JG, Stager JC (1985) Microgeographic diversity in the neotropical knife-fish *Eigenmannia macrops* (Gymnotiformes, Sternopygidae). Environ Biol Fishes 13:173–181

Lutz PL, McMahon PM, Rosenthal M, Sick TJ (1984) Relationship between aerobic and anaerobic energy production in turtle brain in situ. Am J Physiol 247:R740–R744

Lutz PL, Rosenthal M, Sick TJ (1985) Living without oxygen: turtle brain as a model of anaerobic metabolism. Mol Physiol 8:411–425

Machado-Allison A (1971) Contribucion al conocimiento de la taxonomia des genero *Cichla* (Perciformes: Cichlidae) en Venezuela. Part I. Acta Biol Venez 7(4):459–497

Machado-Allison A (1987) Los peces de los llanos de Venezuela. Un ensayo sobre su historia natural. Universidad Central de Venezuela, Caracas, 144 pp

Mago-Leccia F (1978) Los peces de la familia Sternopygidae de Venezuela, incluyendo un descripcion de la osteologia de *Eigenmannia virescens* y una nueva definicion y clasificacion del orden Gymnotiformes. Acta Cient Venez 29(1):1–89

Mago-Leccia F, Lundberg JG, Baskin JN (1985) Systematics of the South American freshwater fish genus *Adontosternarchus* (Gymnotiformes, Apteronotidae). Contrib Sci (Los Ang) 358:1–19

Markert CL (1984) Lactate dehydrogenase – biochemistry and function of lactate dehydrogenase. Cell Biochem Function 2:131–134

Markert CL, Holmes RS (1969) Lactate dehydrogenase of the flatfish, Pleuronectiformes: kinetic, molecular, and immunochemical analysis. J Exp Zool 159:319–332

Markert CL, Shaklee JB, Whitt GS (1975) Evolution of a gene. Science 189:102–114

Marlier G (1967) Ecological studies on some lakes of the Amazon valley. Amazoniana 1(2): 91–115

Marlier G (1968) Les poissons du lac Redondo et leur régime alimentaire; les chaînes trophiques du lac Redondo; les poissons du rio Prêto da Eva. Cadernos da Amazônia 11:21–57

Marlier G (1973) Limnology of the Congo and Amazon rivers. In: Meggers BJ, Ayensu ESA, Duckworth WD (eds) Tropical forest ecosystems in Africa and South America: a comparative review. Smithsonian Inst Press, Washington, DC, pp 223–238

Martin JP, Bonaventura J, Fyhn HJ, Fyhn UEH, Garlick RL, Powers DA (1979a) Structural and functional studies of hemoglobins from Amazon stingrays of the genus Potamotrygon. Comp Biochem Physiol 62A:131–138

Martin JP, Bonaventura J, Brunori M, Fyhn HJ, Fyhn UEH, Garlick RL, Powers DA, Wilson MT (1979b) The isolation and characterization of the hemoglobin components of *Mylossoma* sp., an Amazonian teleost. Comp Biochem Physiol 62A:155–162

Martinelli LA (1986) Composição química e isotópica (^{13}C) de sedimentos de várzea e suas interações com alguns rios da bacia Amazônica. MSc Thesis, Escola Superior "Luiz de Queiroz", Piracicaba, 214 pp

Maués RH (1991) "Amazônias": regional identity and national integration. Ciênc Cult 43(1): 26–31

Mayr E (1977) Populações, espécies e evolução. Editôra Nacional/EDUSP, São Paulo, 485 pp

McMahon BR (1970) The relative efficiency of gaseous exchange across the lungs and gills of an African lungfish *Protopterus aethiopicus*. J Exp Biol 52:1–15
Menezes NA, Vazzoler AEA de M (1992) Reproductive characteristics of characiformes. In: Hamlett WC (ed) Reproductive biology of South American vertebrates. Springer, New York, pp 60–70
Menezes RS (1951) Notas Biológicas e Econômicas Sobre o Pirarucu. Série de Estudos técnicos No 3, Ministério da Agricultura, Rio de Janeiro
Migdalski EC (1957) Contribution to the life history of the South American fish *Arapaima*. Copeia 1957:55–56
Milligan CL, Wood CM (1987) Regulation of blood oxygen transport and red cell pHi after exhaustive activity in rainbow trout (*Salmo gairdneri*) and starry flounder (*Platichthys stellatus*). J Exp Biol 133:263–282
Mommsen TP, French CJ, Hochachka PW (1980) Sites and patterns of protein and amino acid utilization during the spawning migration of salmon. Can J Zool 58:1785–1799
Monteiro MC, Schwantes ML, Schwantes AR (1991) Malate dehydrogenase in subtropical fish belonging to the orders Characiformes, Siluriformes and Perciformes. I. Duplicate gene expression and polymorphism. Comp Biochem Physiol 100B(2):381–389
Monteiro PJC, Val AL, Almeida-Val VMF (1987) Biological aspects of Amazonian fishes: Hemoglobin, hematology, intraerythrocytic phosphates, and whole blood Bohr effect of *Mylossoma duriventris*. Can J Zool 65:1805–1811
Moon TW, Hochachka PW (1971) Temperature and enzyme activity in poikilotherms. Isocitrate dehydrogenase in rainhow trout liver. Biochem J 123:695–705
Moura MAF (1990) Aspectos hematológicos – série branca em peixes da região Amazônica. Internal Rep, CNPq, Manaus, 39 pp
Moyes CD, Mathieu-Costello OA, Brill RW, Hochachka PW (1992) Mitochondrial metabolism of cardiac and skeletal muscles from a fast (*Katsuwonus pelamis*) and a slow (*Cyprinus carpio*) fish. Can J Zool 70:1246–1253
Moyle PB, Cech JJ (1982) Fishes: an introduction to ichthyology. Prentice Hall, Englewood Cliffs
Munro AD (1990) Tropical freshwater fish. In: Munro AD, Scott AP, Lam TJ (eds) Reproductive seasonality in teleosts: environmental influences. CRC Press, Boca Raton, pp 145–188
Murphy JB, Nance RD (1992) Mountain belts and the supercontinent cycle. Sci Am 266(4): 84–91
Nakamoto W, Machado PEA, Foresti F (1986) Hemoglobin patterns in different populations of *Synbranchus marmoratus* Bloch, 1795 (Pisces, Synbranchidae). Comp Biochem Physiol 84B:377–381
Nei M (1987) Molecular evolutionary genetics. Columbia Univ, New York, 512 pp
Nelson JS (1984) Fishes of the world, 2nd edn. Wiley, New York
Nikinmaa M (1990) Vertebrate red blood cells. Adaptations of function to respiratory requirements. Springer, Berlin Heidelberg New York, 262 pp
Nikinmaa M, Tuurula H, Soivio A (1980) Thermoacclimatory changes in blood oxygen binding properties and gill secondary lamellar structure of *Salmo gairdneri*. J Comp Physiol 140: 255–260
Nilsson S, Grove DJ (1975) Adrenergic and cholinergic innervation of the spleen of the cod, *Gadus morhua*. Eur J Pharmacol 28:135–143
Noda SN, Noda H (1990) A água envenenada. Ciênc Hoje 11(64):40
Northcote TG (1984) Mechanisms of fish migration in rivers. In: McCleave JD, Arnold GP, Dodson JJ, Neil WH (eds) Mechanisms of migration in fishes. Proc NATO Adv Res mechanisms of migration in fishes. Plenum Press, New York, pp 317–355
Novacek MJ, Marshall LG (1976) Early biogeographic history of Ostariophysan fishes. Copeia 1976:1–12
Ohno S (1970) The enormous diversity in genome size of fish as a reflection of nature's extensive experiments with gene duplication. Trans Am Fish Soc 99:120–130

Ohno S, Atkin NB (1966) Comparative DNA values and chromosome complement in eight species of fishes. Chromosoma 18:455–456

Oliveira de AE (1983) Ocupação humana. In: Salati E, Junk WJ, Schubart HOR, Oliveira de AE (eds) Amazônia: desenvolvimento, integração e ecologia. Editôra Brasiliense and CNPq, Brasília, pp 144–327

Oliveira C, Almeida-Toledo LF, Toledo-Filho SA (1990) Comparative cytogenetics analysis of three cytotypes of *Corydoras nattereri* (Pisces, Siluriformes, Callichthyidae). Cytologia 55:21–26

Oltman RE (1967) Reconnaissance investigations of the discharge water quality of the Amazon. Atas simpósio sôbre biota Amazônica, Rio de Janeiro, 3(Limnologia):163–185

Packard GC (1974) The evolution of air-breathing in Paleozoic gnathostome fishes. Evolution 28:320–325

Paixão IMP (1980) Estudos da alimentação e reprodução de *Mylossoma duriventris* Cuvier, 1818 (Pisces, Characoidei) do lago Janauacá, AM, Brasil. MSc Thesis, PPG INPA/FUA, Manaus, 127 pp

Panepucci LLL, Schwantes ML, Schwantes AR (1984) Loci that encode the lactate dehydrgenase in 23 species of fish belonging to the orders Cypriniformes, Siluriformes, and Perciformes: adaptative features. Comp Biochem Physiol 77B:867–876

Pauls E, Bertollo LAC (1990) Distribution of a supernumerary chromosome system and aspects of karyotypic evolution in the genus *Prochilodus* (Pisces, Prochilodontidae). Genetica 81: 117–123

Pelster B, Weber RE (1990) Influence of organic phosphates on the Root effect of multiple fish haemoglobins. J Exp Biol 149:425–437

Pelster B, Weber RE (1991) The physiology of the Root effect. In: Advances in comparative environmental physiology, vol 8. Springer, Berlin Heidelberg New York, pp 51–77

Pereira M (1991) Desenvolvimento e preservação das áreas de várzea da Amazônia brasileira. In: Val AL, Feldberg E, Figliuolo R (eds) Bases científicas para estratégias de preservação e desenvolvimento da Amazônia: fatos e perspectivas. Imprensa da Universidade do Amazonas, Manaus, pp 55–63

Pereira M, Guimarães SF, Storti A, Graef EW (1991) Piscicultura na Amazônia Brasileira: entraves ao seu desenvolvimento. In: Val AL, Feldberg E, Figliuolo R (eds) Bases científicas para Estratégias de preservação e desenvolvimento da Amazônia: fatos e perspectivas. Imprensa da Universidade do Amazonas, Manaus, pp 373–380

Perez JE, Rylander MK (1985) Hemoglobin heterogeneity in Venezuelan fishes. Comp Biochem Physiol 80B:641–646

Perry SF, Kinkead R (1989) The role of catecholamines in regulating arterial oxygen content during acute hypercapnic acidosis in rainbow trout (*Salmo gairdneri*). Resp Physiol 77: 365–377

Perry SF, Laurent P (1990) The role of carbonic anhydrase in carbon dioxide excretion, acid-base balance and ionic regulation in aquatic gill breathers. In: Truchot JP, Lahlou B (eds) Animal nutrition and transport processes 2. Transport, respiration and excretion: comparative and environmental aspects. Karger, Basel, pp 39–57

Petrere M Jr (1978a) Pesca e esforço de pesca no Estado do Amazonas. I. Esforço e captura por unidade de esforço. Acta Amazonica 8:439–454

Petrere M Jr (1978b) Pesca e esforço de pesca no Estado do Amazonas. II. Locais, aparelhos de captura e estatísticas de desembarque. Acta Amazonica 8(Suppl 2):54

Petrere M Jr (1983) Relationship among catches, fishing effort and river morphology for eight rivers in Amazonas state (Brazil), during 1976–1978. Amazoniana 7(2):281–296

Petry P (1989) Deriva de ictioplankton no paraná do Rei, várzea do Careiro, Amazônia Central, Brasil. MSc Thesis, PPG INPA/FUA, 68 pp

Phelps C, Farmer M, Fyhn HJ, Fyhn UEH, Garlick RL, Noble RW, Powers DA (1979a) Equilibria and kinetics of oxygen and carbon monoxide ligand binding to the hemoglobin of the South American lugfish, *Lepidosiren paradoxa*. Comp Biochem Physiol 62A(1):139–144

Phelps C, Garlick RL, Powers DA, Noble RW, Martin JP (1979b) Equilibria and kinetics of oxygen and carbon monoxide binding to the hemoglobin of the teleost *Synbranchus marmoratus*. Comp Biochem Physiol 62A:227–230

Phleger CF, Saunders BS (1978) Swim-bladder surfactants of Amazon air-breathing fishes. Can J Zool 56:946–952

Pires JM, Prance GT (1985) The vegetation types of the Brazilian Amazon. In: Prance GT, Lovejoy TE (eds) Amazonia. Key environments. Pergamon Press, Oxford, pp 109–145

Porto JIR (1992) Estudos citotaxonômicos em peixes da família Hemiodidae (Ostariophysi, Characiformes), da Amazônia central. MSc Thesis, PPG INPA/FUA, Manaus, 103 pp

Porto JIR, Feldberg E (1992) Comparative cytogenetic study of the armoured catfishes of the genus *Hoplosternum* (Siluriformes, Callichthyidae). Brazil. J Genet 15(2):359–367

Porto JIR, Feldberg E, Nakayama CM, Maia RO, Jégu M (1991) Cytotaxonomic analysis in the Serrasalmidae (Ostariophysi, Chraciformes). VII Int Icthyoiogy Congr, the Hague, Abstract Book, p 66

Porto JIR, Feldberg E, Nakayama CM, Falcão JN (1992) A checklist of chromosome numbers and karyotypes of Amazonian freshwater fishes. Rev Hydeobiol Trop 25(4):287–299

Powers DA (1972) Hemoglobin adaptation for fast and slow water habitats in sympatric catostomid fishes. Science 177:360–362

Powers DA (1974) Structure, function and molecular ecology of fish hemoglobins. Ann NY Acad Sci 241:472–490

Powers DA (1980) Molecular ecology of teleost fish hemoglobins: strategies for adapting to changing environments. Am Zool 20:139–162

Powers DA, Martin JP, Garlick RL, Fyhn HJ, Fyhn UEH (1979) The effect of temperature on the oxygen equilibria of fish hemoglobins in relation to environmental thermal variability. Comp Biochem Physiol 62A:87–94

Precht H, Christophersen J, Hensel H, Larcher W (1973) Temperature and life. Springer, Berlin Heidelberg New York, 779 pp

Primmett DRN, Randall DJ, Mazeaud M, Boutilier RG (1986) The role of catecholamines in erythrocyte pH regulation and oxygen transport in rainbow trout (*Salmo gairdneri*) during exercise. J Exp Biol 122:139–148

Pritchard AW, Hunter JR, Lasker R (1974) The relation between exercise and biochemical changes in red and white muscle and liver in the jack mackerel, *Trachurus symmetricus*. Fish Res Bull 69:379–386

Prosser CL (1991) Temperature. In: Prosser CL (ed) Environmental and metabolic animal physiology. Wiley, New York, pp 109–165

Putzer H (1976) Metallogenetische Provinzen in Südamerika. Schweizerbart, Stuttgart, 316 pp

Putzer H (1984) The geological evolution of the Amazon basin and its mineral resources. In: Sioli H (ed) The Amazon. Limnology and landscape ecology of a mighty tropical river and its basin. Junk, Dordrecht, pp 15–46

Rahn H, Howell BJ (1976) Bimodal gas exchange. In: Hughes GM (ed) Respiration of amphibious vertebrates. Academic Press, New York, pp 271–285

Rahn H, Rahn KB, Howell BJ, Gans C, Tenney SM (1971) Air breathing of the garfish (*Lepisosteus osseus*). Respir Physiol 11:285–307

Rai H, Hill G (1984a) Primary production in the Amazonian aquatic ecosystem. In: Sioli H (ed) The Amazon. Limnology and landscape ecology of a mighty tropical river and its basin. Junk, Dordrecht, pp 311–336

Rai H, Hill G (1984b) Microbiology of Amazonian waters. In: Sioli H (ed) The Amazon. Limnology and landscape ecology of a mighty tropical river and its basin. Junk, Dordrecht, pp 413–442

Rambhaskar B, Rao K (1987) Comparative hematology of ten species of marine fish from Visakhapatnam coast. J Fish Biol 30:59–66

Ramirez-Gil H (1993) Características genéticas, moleculares e fisiológicas de *Callophysus macropterus* (Siluriformes, Pimelodidae) do rio Solimões e do rio Negro (Amazônia Central). MSc Thesis, PPG INPA/FUA, Manaus, 109 pp

Randall DJ (1982) The control of respiration and circulation in fish during exercise and hypoxia. J Exp Biol 100:275–288
Randall DJ (1990) Control and co-ordination of gas exchange in water breathers. In: Boutilier RG (ed) Advances in comparative and environmental physiology. Vertebrate gas exchange: from environment to cell. Springer, Berlin Heidelberg New York, pp 253–278
Randall DJ, Perry SF (1992) Catecholamines. In: Hoar WS, Randall DJ, Farrel AP (eds) Fish physiology, vol XIIB. Academic Press, New York, pp 255–300
Randall DJ, Val AL (1993) Carbon dioxide and acid transfer in teleost fish. In: Scheid P (ed) Proc Symp "Respiration in health and disease: lessons from comparative physiology". Gustav Fischer, Stuttgart (in press)
Randall DJ, Val AL. The role of carbonic anhydrase in aquatic gas exchange. In: Heisler N (ed) Advances in comparative and environmental physiology. Mechanisms of systemic regulation in lower vertebrates: respiration and circulation. Springer, Berlin Heidelberg New York (in press)
Randall DJ, Farrell AP, Haswell MS (1978a) Carbon dioxide excretion in the pirarucu (*Arapaima gigas*), an obligate air-breathing fish. Can J Zool 56:977–982
Randall DJ, Farrell AP, Haswell MS (1978b) Carbon dioxide excretion in the jejú (*Hoplerythrinus unitaeniatus*), a facultative air-breathing fish. Can J Zool 56:970–973
Randall DJ, Burggren WW, Farrell AP, Haswell MS (1981) The evolution of air-breathing vertebrates. Cambridge Univ Press, Cambridge, 133 pp
Rantin FT, Johansen K (1984) Responses of the teleost *Hoplias malabaricus* to hypoxia. Env Biol Fish 11:221–228
Rapp-Py-Daniel LH, Leão ELM (1991) A coleção de peixes do INPA: base do conhecimento científico sobre a ictiofauna Amazônica gerado pelo Instituto Nacional de Pesquisas da Amazônia. In: Val AL, Feldberg E, Figliuolo R (eds) Bases científicas para estratégias de preservação e desenvolvimento da Amazônia: fatos e perspectivas. Imprensa da Universidade do Amazonas, Manaus, pp 299–312
Raven PH, Johnson GB (1986) Biology. Times Mirror/Mosby College, St Louis
Reeves RB (1977) The interaction of body temperature and acid-base balance in ectothermic vertebrates. Ann Rev Physiol 39:559–586
Reichlin M, Davis BJ (1979) Antigenic relationships among fishes common to the Amazon river basin. Comp Biochem Physiol 62A:101–104
Reischl E, Tondo CV (1974) Multiple hemoglobins in fish. II. The sub-unit composition of the hemoglobins of the fresh-water *Pimelodus maculatus*: detection of asymmetric tetramers. Rev Bras Biol 34:337–344
Renno JF, Berrebi P, Boujard T, Guyomard (1990) Intraspecific genetic differentiation of *Leporinus friderici* (Anostomidae: Pisces) in French Guiana and Brazil: a genetic approach to the refuge theory. J Fish Biol 36:85–95
Ribeiro JSB (1978) Fatores ecológicos, produção primária e fitoplancton em cinco lagos da Amazônia Central. Estudo preliminar. MSc Thesis, PPG ERN, São Carlos, 143 pp
Ribeiro JSB, Darwich AJ (1993) Produção primária fitoplanctônica de um lago de ilha fluvial na Amazônia Central (Lago do Rei, Ilha do Careiro). Amazoniana XII (3/4):365–383
Ribeiro MCL (1983) As migrações dos jaraquis (Pisces, Prochilodontidae) no rio Negro, Amazonas, Brasil. MSc Thesis, PPG INPA/FUA, Manaus, 192 pp
Ribeiro MCL (1985) A natural hybrid between two tropical fishes: *Semaprochilodus insignis* × *Semaprochilodus taeniurus* (Teleostei, Characoidei, Prochilodontidae). Rev Bras Biol 2(7): 419–421
Ribeiro MNG, Salati E, Villa Nova NA, Demetrio CGB (1982) Radiação solar disponível em Manaus e sua relação com a duração do brilho solar. Acta Amazonica 12:339–346
Richey JE, Hedges JI, Devol AH, Quay PD, Victoria R, Martinelli L, Forsberg BR (1990) Biogeochemistry of carbon in the Amazon river. Limnol Oceanogr 35:352–371
Riggs A (1976) Factors in the evolution of hemoglobin function. Fed Am Soc Exp Biol 35:2115–2118

Riggs A (1979) Studies of the hemoglobins of Amazonian fishes: an overview. Comp Biochem Physiol 62A:257–272

Riggs A, Fyhn HJ, Fyhn UEH, Noble RW (1979) Studies of the functional properties of the hemoglobins of *Hoplias malabaricus* and *Hoplerythrinus unitaeniatus*. Comp Biochem Physiol 62A:189–194

Roberts TR (1972) Ecology of the fishes in the Amazon and Congo basins. Bull Mus Comp Zool 143(2):117–147

Robertson BA, Hardy ER (1984) Zooplankton of Amazonian lakes and rivers. In: Sioli H (ed) The Amazon. Limnology and landscape ecology of a mighty tropical river and its basin. Junk, Dordrecht, pp 337–352

Rocha VJ, Giuliano-Caetano L (1990) Considerações cariotípicas na família Hemiodontidae (Characiformes, Teleostei). Resumos do III Simp Citog Evol e Aplic Peixes Neotrop (Abstr):19

Rodewald K, Stangl A, Braunitzer G (1984) Primary structure, biochemical and physiological aspects of hemoglobin from the South American lungfish (*Lepidosiren paradoxa*, Dipnoi). Hoppe-Seyler's Z Physiol Chem 365:639–649

Rooney CHT, Ferguson A (1985) Lactate dehydrogenase isozymes and allozymes of the nine-spined stickleback *Pungitius pungitius* (L.) (Osteichthyes: Gasterosteidae). Comp Biochem Physiol 81B:7111–715

Root RW (1931) The respiratory function of the blood of marine fishes. Biol Bull Mar Biol Lab Woods Hole 61:427–456

Rowley AF, Hunt TC, Page M, Mainwaring G (1988) Fish. In: Rowley AF, Ratcliffe NA (eds) Vertebrate blood cells. Cambridge Univ Press, Cambridge, pp 19–128

Russel CE (1987) Plantation forestry. The Jari Project, Pará, Brazil. In: Jordan CF (ed) Amazonian rain forests. Ecosystem disturbance and recovery. Springer, Berlin Heidelberg New york, pp 76–89

Saint-Paul U (1983) Investigations on the respiration of the Neotropical fish, *Colossoma macropomum* (Serrasalmidae). The influence of weight and temperature on the routine oxygen consumption. Amazoniana 7:433–444

Saint-Paul U, Werder U, Teixeira AS (1981) O uso do aguapé (*Eichornia crassipes*) em experimentos de alimentação de matrinchã (*Brycon* sp.). Ann II Simp Bras Aquic, pp 83–84

Salama A, Nikinmaa M (1988) The adrenergic responses of carp (*Cyprinus carpio*) red cells: effects of PO_2 and pH. J Exp Biol 136:405–416

Salati E (1985) The climatology and hydrology of Amazonia. In: Prance GT, Lovejoy TE (eds) Amazonia. Key environments. Pergamon Press, Oxford, pp 18–48

Salati E, Marques J (1984) Climatology of the Amazon region. In: Sioli H (ed) The Amazon. Limnology and landscape ecology of a mighty tropical river and its basin. Junk, Dordrecht, pp 85–126

Salati E, Vose PB (1984) Amazon basin: A system in equilibrium. Science 225:129–138

Salati E, Ribeiro MNG, Absy ML, Nelson BW (1991) Clima da Amazônia: Presente, Passado e Futuro. In: Val AL, Feldberg E, Figliuolo R (eds) Bases científicas para estratégias de preservação e desenvolvimento da Amazônia: fatos e perspectivas. Imprensa da Universidade do Amazonas, Manaus, pp 21–34

Saldarriaga JG (1987) Recovery following shifting cultivation. A century of sucession in the upper rio Negro. In: Jordan CF (ed) Amazonian rain forests. Ecosystem disturbance and recovery. Springer, Berlin Heidelberg New York, pp 24–33

Salvo-Souza RH (1990) Parâmetros hematológicos (série vermelha) de *Arapaima gigas* (Osteoglossiformes) durante o primeiro ano de vida em cativeiro. MSc Thesis, PPG INPA/FUA, Manaus, 99 pp

Salvo-Souza RH, Val AL (1990) O gigante das águas doces. Ciênc Hoje 11(64):9–12

Santos GM (1980a) Estudo da reprodução e hábitos reprodutivos de *Schizodon fasciatus*, *Rhytiodus microlepis* e *Rhytiodus argenteofuscus* (Pisces, Anostomidae) do lago Janauaca. Acta Amazonica 10(2):391–400

Santos GM (1980b) Aspectos da sistemática e morfologia de *Schizodon fasciatus* Agassiz, 1829, *Rhytiodus microlepis* Kner, 1859 e *Rhytiodus argenteofuscus* Kner, 1859 (Osteichthyes, Characoidei, Anostomidae) do lago Janauaca, Amazonas. Acta Amazonica 10(3): 635–649

Santos GM (1982) Caracterização, hábitos alimentares e reprodutivos de quatro espécies de "aracús" e considerações ecológicas sobre o grupo no lago Janauacá, AM (Osteichthyes, Characoidei, Anostomidae). Acta Amazonica 12(4):713–739

Santos GM (1991) Pesca e ecologia dos peixes de Rondônia. PhD Thesis, PPG INPA/FUA, Manaus, 213 pp

Santos GM, Jegu M (1989) Inventário taxonômico e redescrição das espécies de Anostomídeos (Characiformes, Anostomidae) do baixo rio Tocantins, PA, Brasil. Acta Amazonica 19: 159–213

Santos GM, Carvalho FM (1982) Levantamento preliminar, pesca e aspectos biológicos da ictiofauna do rio Araguaia. Relatório do Projeto Santa Izabel. Convênio Eletronorte/CNPq-INPA, Manaus, 58 pp

Santos GM, Jegu M, Merona B (1984) Catálogo dos peixes comerciais do baixo rio Tocantins. Convênio Eletronorte/CNPq-INPA, Manaus, 83 pp

Santos GM, Ferreira EJG, Zuanon JA (1991) Ecologia de peixes da Amazônia. In: Val AL, Feldberg E, Figliuolo R (eds) Bases científicas para estratégias de preservação e desenvolvimento da Amazônia: fatos e perspectivas. Imprensa da Universidade do Amazonas, Manaus, pp 263–280

Sastry KV, Rao DR (1981) Enzymological and biochemical changes produced by mercuric chloride on the activities of some enzymes in certain tissues of the freshwater murrel, *Channa punctatus*. Toxicol Lett 9:321–326

Satchell GH (1991) Physiology and form of fish circulation. Cambridge Univ Press, Cambridge, 235 pp

Sawaya P (1946) Sobre a biologia de alguns peixes de respiração aérea (*Lepidosiren paradoxa* Fitzinger e *Arapaima gigas* Cuvier). Bol Fac Filos Cienc Let Univ Sao Paulo II:255–286

Scheel JJ (1973) Fish chromosome and their evolution. Internal reports of Danmarks Akvarium, Charlottenlund, Denmark, 22 pp

Schmidt GW (1973) Primary production of phytoplankton in the three types of Amazonian waters. II. The limnology of a tropical floodplain lake in central Amazonia (Lago Castanho). Amazoniana 4(2):139–203

Schmidt-Nielsen K (1984) Scaling. Why is animal size so important? Cambridge Univ Press, New York, 241 pp

Scholander PF, Flagg W, Walters W, Irving L (1953) Climatic adaptation in arctic and tropical poikilotherms. Physiol Zool 26:67–92

Scholander PV, van Dam L (1954) Secretion of gases against high pressures in the swimbladders of deep sea fishes. I. Oxygen dissociation in blood. Biol Bull Mar Biol Lab Woods Hole 107:247–259

Schubart HOR (1983) Ecologia e utilização das florestas. In: Salati E, Junk WJ, Schubart HOR, Oliveira de AE (eds) Amazônia: desenvolvimento, integração e ecologia. Editôra Brasiliense and CNPq, pp 101–143

Schwantes MLB, Schwantes AR (1982) Adaptative features of ectothermic enzymes. I. Temperature effects on the malate dehydrogenase from a temperate fish *Leiostomus xanthurus*. Comp Biochem Physiol 72B:49–58

Schwartz JM, Reiss AL, Jaffe ER (1983) Hereditary methemoglobinemia with deficiency of NADH cytochrome b5 reductase. In: Stanbury JB (ed) Metabolic basis of inherited disease, 5th edn. McGraw-Hill, New York, pp 1654–1655

Schwassmann HO (1976) Ecology and taxonomic status of different geographic populations of *Gymnorhamphichthys hypostomus* Ellis (Pisces, Cypriniformes, Gymnotoidei). Biotropica 8:25–40

Schwassmann HO (1978) Times of annual spawning and reproductive strategies in Amazonian fishes. In: Thorpe JE (ed) Rhythimic activity in fishes. Academic Press, New York, pp 187–200

Sensabaugh GE Jr, Kaplan NO (1972) A lactate dehydrogenase specific to the liver of gadoid fish. J Biol Chem 217:585–593

Shaklee JB, Whitt GS (1981) Lactate dehydrogenase isozymes of gadiform fishs: divergent patterns of gene expression indicate a heterogeneous taxon.

Shaklee JB, Christiansen JA, Sidell BD, Prosser CL, Whitt GS (1977) Molecular aspects of temperature acclimation in fish: contributions of changes in enzymes patterns to metabolic reorganization in the green sunfish. J Exp Zool 201:1–20

Shaklee JB, Allendorf FW, Morizot DC, Whitt GS (1989) Genetic nomenclature for protein-coding loci in fish: proposed guidelines. Trans Am Fish Soc 118:218–227

Shoubridge EA (1980) The metabolic strategy of the anoxic goldfish. PhD Thesis Univ British Columbia, Vancouver, 167 pp

Shoubridge EA, Hochachka PW (1980) Ethanol: novel end product of vertebrate anaerobic metabolism. Science 209:308

Shukla J, Nobre CA, Sellers P (1990) Amazonia deforestation and climate change. Science 247:1322–1325

Sidell BD, Beland KF (1980) Lactate dehydrogenases of Atlantic hagfish: physiological and evolutionary implications of a primitive heart isozyme. Science 207:769–770

Silva JWB (1988) A aquicultura nas regiões norte e nordeste do Brasil. Cong Bras Eng Pesca, pp 24–49

Singh BN (1976) Balance between aquatic and aerial respiration. In: Hughes GH (ed) Respiration of amphibious vertebrates. Academic Press, New York, pp 125–164

Singh BN, Hughes GM (1971) Respiration of an air-breathing catfish *Clarias batrachus* (Linn.) J Exp Biol 55:421–434

Sioli H (1984a) Introduction: history of the discovery of the Amazon and research of Amazonian waters and landscapes. In: Sioli H (ed) The Amazon. Limnology and landscape ecology of a mighty tropical river and its basin. Junk, Dordrecht, pp 1–14

Sioli H (1984b) The Amazon and its main afluents: hydrography, morphology of the river courses, and river types. In: Sioli H (ed) The Amazon. Limnology and landscape ecology of a mighty tropical river and its basin. Junk, Dordrecht, pp 127–166

Sioli H (1984c) Present "development" of Amazonia in the light of the ecological aspect of the life and alternative concepts. In: Sioli H (ed) The Amazon. Limnology and landscape ecology of a mighty tropical river and its basin. Junk, Dordrecht, pp 737–747

Sioli H (1991) Amazônia. Fundamentos da ecologia da maior região de florestas tropicais. Trad J Becker, Editõra Vozes, Petrópolis, 72 pp

Slobodkin LB, Rapoport A (1974) An optimal strategy of evolution. Q Rev Biol 49(3):181–200

Smith DG, Gannon BJ (1978) Selective control of branchial arch perfusion in an air-breathing Amazonian fish *Hoplerythrinus unitaeniatus*. Can J Zool 56:959–964

Smith HM (1930) Metabolism of the lungfish *Protopterus aethiopicus*. J Biol Chem 88:97–130

Smith RJF (1985) The control of fish mighration. Springer, Berlin Heidelberg New Yrok, 243 pp

Soares MGM (1979) Aspectos ecológicos (alimentação e reprodução) dos peixes do igarapé do Porto, Aripuanã, MT. Acta Amazonica 9(2):325–352

Soares MGM (1993) Estratégias respiratórias dos peixes do lago Camaleão (Ilha da Marchantaria) – AM, Br. PhD Thesis, PPG INPA/FUA, Manaus, 175 pp

Soivio A, Nikinmaa M, Westman K (1980) The blood oxygen binding properties of hypoxic *Salmo gairdneri*. J Comp Physiol B 136:83–87

Sola L, Cataudella S, Capanna E (1981) New developments in vertebrate cytotaxonomy. III. Karyology of bony fishes: a review. Genetica 54:285–328

Somero GN (1969) Enzyme mechanisms of temperature compensation: immediate and evolutionary effects of temperatures on enzymes of aquatic poikilotherms. Am Nat 103: 517–530

Somero GN (1975) The roles of isozymes in adaptation to varying temperatures. In: Markert CL (ed) Isozymes. II. Physiological function. Academic Press, New York, pp 221–234

Somero GN (1981) pH-temperature interactions on proteins: principles of optimal pH and buffer system design. Mar Biol Lett 2:163–178
Somero GN (1983) Environmental adaptation of proteins: strategies for the conservation of critical functional and structural traits. Comp Biochem Physiol 76A:621–633
Somero GN, Hochachka PW (1968) The effect of temperature on catalytic and regulatory properties of pyruvate kinases from rainbow trout and the Antartic fish *Trematomus bernacchii*. Biochem J 110:395–400
Souza RHS, Val AL (1990) O gigante das águas doces. Ciênc Hoje 11(64):9–12
Stacey NE (1984) Control of the timing of ovulation by exogenous and endogenous factors. In: Potts GW, Wotton RJ (eds) Fish reproduction: strategies and tactics. Academic Press, New York, pp 207–222
Stanley SM (1990) The general correlation between rate of speciation and rate of extinction: fortuitous causal linkage. In: Ross RM, Allmon WD (eds) Causes of evolution. A paleontological perspective. Univ Chicago Press, Chicago, pp 103–127
Starck D (1978) Vergleichende Anatomie der Wirbeltiere auf evolutionsbiologischer Grundlage, vol 1. Springer, Berlin Heidelberg New York
Stark N (1970) The nutrient content of plants and soils from Brazil and Surinam. Biotropica 2(1):51–60
Steen JB (1963) The physiology of the swimbladder in the eel *Anguilla vulgaris*. III. The mechanism of gas secretion. Acta Physiol Scand 59:221–241
Steen JB, Berg T (1966) The gills of two species of hemoglobin-free fishes compared to those of other teleosts – with a note on severe anaemia in an eel. Comp Biochem Physiol 18: 517–526
Stevens ED, Holeton GF (1978a) The partioning of oxygen uptake from air and from water by the large obligate air-breathing teleost pirarucu (*Arapaima gigas*). Can J Zool 56:974–976
Stevens ED, Holeton GF (1978b) The partioning of oxygen uptake from air and from water by erythrinids. Can J Zool 56:965–969
Stock DW (1991) A phylogenetic analysis of the 18S ribosomal RNA sequence of the coelacanth *Latimeria chalumnae*. Environ Biol Fishes 32:99–117
Storey KB, Guderley HE, Guppy M, Hochachka PW (1978) Control of ammoniagenesis in the kidney of water- and air-breathing osteoglossids: characterization of glutamate dehydrogenase. Can J Zool 56:845–851
Teixeira AS, Jamieson A (1985) Genetic variation in plasma transferrins of tambaqui, *Colossoma macropomum* (Cuvier, 1818). Amazoniana IX(2):159–168
Tetens V, Lykkeboe G (1981) Blood respiratory properties of rainbow trout *Salmo gairdneri*: responses to hypoxia acclimation and anoxic incubation of blood in vitro. J Comp Physiol B 145:117–125
Tetens V, Wells RMG (1984) Oxygen binding properties of blood and hemoglobin solutions in the carpet shark (*Cephaloscyllium isabella*): roles of ATP and urea. Comp Biochem Physiol 79A:165–168
Tetens V, Lykkeboe G, Christiansen NJ (1988) Potency of adrenaline and noradrenaline for β-adrenergic proton extrusion from red cells of rainbow trout, *Salmo gairdneri*. J Exp Biol 134:267–280
Thompson KW (1979) Cytotaxonomy of 41 species of Neotropical Cichlidae. Copeia 4:679–691
Thorgaard GH (1983) Chromosomal differences among rainbow trout populations. Copeia 1983:650–662
Thorson TB (1972) The status of the bull shark *Carcharhinus leucas*, in the Amazon river. Copeia 1972:601–605
Treece D (1989) Brazil's greater Carajás program. In: Schneider A (ed) Deforestation and "development" in Canada and the Tropics. The impact on people and environment. Univ Cape Breton, Sydney, 28 pp
Turner BJ, Diffoot N, Rasch EM (1992) The callichthyid catfish *Corydoras aeneus* is an unresolved diploid-tetraploid sibling species complex. Ichthyol Explor Freshwaters 3(1): 17–23

Tyler SC, Blake DR, Rowland FS (1987) $^{13}C/^{12}C$ ratio in the methane from the flooded Amazon forest. J Geophys Res 92:1044–1048
Uhl C, Buschbacher R (1991) Queimada: O corte que atrai. Ciênc Hoje 7(40):24–28
Val AL (1983) Aspectos estruturais e funcionais de hemoglobinas de espécies do gênero *Semaprochilodus* (Prochilodontidae), do rio Negro, AM, Brasil. MSc Thesis, PPG INPA/FUA, Manaus, 192 pp
Val AL (1986) Hemoglobinas de *Colossoma macropomum* Cuvier, 1818 (Characoidei, Pisces): aspectos adaptativos (Ilha da Marchantaria, Manaus, AM). PhD Thesis, PPG INPA/FUA, Manaus, 112 pp
Val AL (1993) Adaptations of fish to extreme conditions in fresh waters. In: Bicudo JE (ed) The vertebrate gas transport cascade: adaptations to environment and mode of life. CRC Press, Boca Raton, pp 43–53
Val AL, Almeida-Val VMF (1988) Adaptative features of Amazon fishes. Hemoglobins of *Brycon* cf. *cephalus* and *Brycon* cf. *erythropterum*. Rev Bras Genet 11:27–39
Val AL, Almeida-Val VMF, Schwantes AR, Schwantes ML (1984) Biological aspects of Amazonian fishes. I. Red blood cell phosphates of schooling fishes (genus *Semaprochilodus*: Prochilodontidae). Comp Biochem Physiol 78B:215–217
Val AL, Schwantes AR, Almeida-Val VMF, Schwantes MLB (1985) Hemoglobin, hematology, intraerythrocytic phosphates and whole blood Bohr effect from lotic and lentic *Hypostomus regani* populations (São Paulo-Brasil). Comp Biochem Physiol 80B:737–741
Val AL, Schwantes AR, Almeida-Val VMF (1986) Biological aspects of Amazonian fishes-VI. Hemoglobins and whole blood properties of *Semaprochilodus* species (Prochilodontidae) at two phases of migration. Comp Biochem Physiol 83B(3):659–667
Val AL, Almeida-Val VMF, Monteiro PJC (1987) Aspectos biológicos de peixes amazônicos. IV. Padrões eletroforéticos de hemoglobinas de 22 espécies coletadas na Ilha da Marchantaria (Manaus-AM). Acta Amazonica 16/17:125–134
Val AL, Almeida-Val VMF, Affonso EG (1990) Adaptative features of Amazon fishes: hemoglobins, hematology, intraerythrocytic phosphates and whole blood Bohr effect of *Pterygoplichthys multiradiatus*. Comp Biochem Physiol 97B:435–440
Val AL, Affonso EG, Almeida-Val VMF (1992a) Adaptative features of Amazon fishes: blood characteristics of Curimatã (*Prochilodus cf nigricans*, Osteichthyes). Physiol Zool 65(4): 832–843
Val AL, Affonso EG, Souza RHS, Almeida-Val VMF, Moura MAF (1992b) Inositol pentaphosphate in the erythrocytes of an Amazonian fish, the pirarucú (*Arapaima gigas*). Can J Zool 70:852–855
Val AL, Mazur CF, Salvo-Souza RH, Iwama GK (1994) Effects of experimental anaemia on intraerythrocytic phosphate levels in rainbow trout (*Oncorhynchus mykiss*). J Fish Biol 45:269–277
Van der Bank FH, Grant WS, Ferreira JT (1989) Electrophoretically detectable genetic data for fifteen mitochondrial malate dehydrogenase isozymes in the teleosts. Experientia 26: 734–736
Van Waarde A, Van den Thillart G, Verhagen M (1993) Ethanol formation and pH-regulation in fish. In: Hochachka PW, Thillart Van den G, Lutz P (eds) Surviving hypoxia: mechanisms of control versus adaptation. CRC Press, Boca Raton, pp 157–170
Vanzolini PE (1973) Paleoclimates, relief and species multiplication in equatorial forest. In: Meggers BJ, Agensu E, Duckworth R (eds) Tropical forest ecosystems in Africa and South America. Smithsonian Inst Press, Washington, DC, pp 225–258
Vari RP (1989) A phylogenetic study of the neotropical characiform family Curimatidae (Pisces: Ostariophysi). Smithson Contrib Zool 471:69
Vazzoler AEA de M (1983) Padrões eletroforéticos de proteínas totais de cristalinos de espécies do gênero *Semaprochilodus* da bacia Amazônica. Ciênc Cult 35(7):528
Vazzoler AEA de M (1992) Estado atual do conhecimento da Ictiologia no Brasil. Reprodução de Peixes. In: Agostinho AA, Cecílio EB (eds) Situação atual e perspectivas da Ictiologia no Brasil. Editôra da Universidade de Maringá, Maringá, PR, pp 1–13

Vazzoler AEA de M, Caraciolo-Malta MC, Braga FM (1983) Caracterização morfológica do gênero *Semaprochilodus* da bacia Amazônica: caracteres merísticos. Ciênc Cult 35(Suppl 7):528

Vazzoler AEA de M, Amadio SA, Caraciolo-Malta MC (1989a) Aspectos biológicos de peixes amazônicos. XI. Reprodução das espécies do gênero *Semaprochilodus* (Characiformes, Prochilodontidae) no baixo rio Negro, Amazonas, Brasil. Rev Bras Biol 49(1):165–175

Vazzoler AEA de M, Amadio SA, Caraciolo-Malta MC (1989b) Aspectos biológicos de peixes amazônicos. XII. Indicadores quantitativos de deseova das espécies do gênero *Semaprochilodus* (Characiformes, Prochilodontidae) do baixo rio Negro, Amazonas, Brasil. Rev Bras Biol 49(1):175–181

Venere PC, Galetti PM Jr (1989) Chromosome relationships of some neotropical characiformes of the family Curimatidae, Brazil. J Genet 12(1):17–25

Victória RL, Brown IF, Martinelli LA, Salati E (1991) A Amazônia brasileira e seu papel no aumento da concentraçąo de CO_2 na atmosfera. In: Val AL, Feldberg E, Figliuolo R (eds) Bases científicas para estratégias de preservação e desenvolvimento da Amazônia: fatos e perspectivas. Imprensa da Universidade do Amazonas, Manaus, pp 9–20

Vieira I (1982) Aspectos sinecológicos da ictiofauna de Curuá-Una, represa hidrelétrica da região Amazônica. Thesis, Univ Juiz de Fora, MG, 104 pp

Walker I (1990) Ecologia e Biologia dos igapós e igarapés. Ciênc Hoje 11(64):44–53

Walker I (1991) Algumas considerações sobre um programa de zoneamento para a Amazônia. In: Val AL, Feldberg E, Figliuolo R (eds) Bases científicas para estratégias de preservação e desenvolvimento da Amazônia: fatos e perspectivas. Imprensa da Universidade do Amazonas, Manaus, pp 37–46

Walsh PJ, Mommsen TP, Moon TW, Perry SF (1988) Effects of acid-base variables on in vitro hepatic metabolism in rainbow trout. J Exp Biol 135:231–241

Weber RE, Lykkeboe G (1978) Respiratory adaptations in carp blood. Influences of hypoxia, red cell organic phosphates, divalent cations and CO_2 on hemoglobin-oxygen affinity. J Comp Physiol 128:127–137

Weber RE, Wood SC (1979) Effects of erythrocyte nucleoside triphosphates on oxygen equilibria of composite and fractioned hemoglobins from the facultative air-breathing Amazonian catfish, *Hypostomus* and *Pterygoplichthys*. Comp Biochem Physiol 62A:179–184

Weber RE, Wood SC, Davis BJ (1979) Acclimation to hypoxic water in facultative air-breathing fish: blood oxygen affinity and allosteric effectors. Comp Biochem Physiol 62A:125–129

Weitzman SH (1962) The osteology of *Brycon meeki*, a generalized characid fish, with an osteological definition of the family. Stanford Ichthyol Bull 8(1):3–77

Weitzman SH, Weitzman M (1982) Biogeography and evolutionary diversification in neotropical freshwater fishes, with comments on the refuge theory. In: Prance GT (ed) Biological diversification in the Tropics. Columbia Univ Press, New York, pp 403–422

Welcome RL (1979) Fisheries ecology of the floodplain rivers. Chancer Press Ltd. Bungay, Suffolk, 317 pp

Wells RMG, Weber RE (1991) Is there an optimal haematocrit for rainbow trout, *Oncorhynchus mykiss* (Walbaum)? An interpretation of recent data based on blood viscosity measurements. J Fish Biol 38:53–65

Weninger G (1985) Principal fresh-water types and comparactive hydrochemistry of tropical running water systems. Rev Hydrobiol Trop 18(2):79–110

West JB (1991) Acclimatization and adaptation: organ to cell. In: Lahiri S, Cherniak NS, Fitzgerald RS (eds) Responses and adaptation to hypoxia: organ to organelle. Oxford Univ Press, New York, pp 177–190

White MJD (1968) Models of speciation. Science 159:1065–1070

White MJD (1978) Chain processes in chromosomal speciation. Syst Zool 27:285–298

Whitmore DH, Goldberg I (1972) Trout intestinal alkaline phosphatases. I. Some physical-chemical characteristics. J Exp Zool 182:47–58

Whitt GS (1975) A unique lactate dehydrogenase isozyme in teleost retina. In: Ali MA (ed) Vision in fish. Plenum, New York, pp 459–470
Whitt GS (1981) Evolution of isozymes loci and their differential regulation. In: Scudder GGE, Reveal JL (eds) Proceedings of the second international congress of systematic and evolutionary biology: evolution today. Univ City Press, Vancouver, pp 271–289
Whitt GS (1983) Isozymes as probes and participants in developmental and evolutionary genetics. In: Ratazzi MC, Scandalios JG, Whitt GS (eds) Isozymes: current topics in biological and medical research, vol 10. Alan R Liss, New York, pp 1–40
Whitt GS (1987) Species differences in isozyme tissue patterns: their utility for systematic and evolutionary analyses. In: Ratazzi MC, Scandalios JG, Whitt GS (eds) Isozymes: Current topics in biological and medical research, vol 15. Alan R Liss, New York, pp 2–23
Whitt GS, Miller ET, Shaklee JB (1973) Developmental and biochemical genetics of lactate dehydrogenase isozymes in fishes. In: Schroeder JH (ed) Genetics and mutagenesis of fishes. Springer, Berlin Heidelberg, New York, pp 243–276
Wilhelm D, Reischl E (1981) Heterogeneity and functional properties of hemoglobins from south Brazilian freshwater fish. Comp Biochem Physiol 69B:463–470
Wilson AC (1975) Evolutionary importance of gene regulation. Stadler Symp 7. Univ Missouri, Columbia, pp 117–134
Wilson AC (1985) The molecular basis of evolution. Sci Am 253:164–173
Wilson AC, Kaplan NO, Levine L, Pesce A, Reichlin M, Allinson WS (1964) Evolution of lactate dehydrogenase. FASEB Fed Proc 23:1258–1266
Wilson FR, Whitt GS, Prosser CL (1973) Lactate dehydrogenase and malate dehydrogenase isozyme patterns in tissues of temperature-acclimated goldfish (*Carassius auratus* L.). Comp Biochem Physiol 46B:105–116
Wilson TL (1977) Interrelations between pH and temperature for the catalytic activity rate of the M_4 isozyme of lactate dehydrogenase (E.C. 1.1.1.27) from goldfish (*Carassius auratus*, L.). Arch Biochem Biophys 179:378–390
Wittenberg JB, Wittenberg BA (1974) The choroid rete mirabile of the fish eye. I. Oxygen secretion and structure: comparison with the swimbladder rete mirabile. Biol Bull 146: 116–136
Wood CM (1976) Pharmacological properties of the adrenergic receptors regulating systemic vascular resistence in the rainbow trout. J Comp Physiol B 107:211–228
Wood SC, Johansen K (1973) Blood oxygen transport and acid-base balance in eels during hypoxia. Am J Physiol 225:849–851
Wood CM, McMahon BR, McDonanald DG (1979a) Respiratory, ventilatory, and cardiovascular responses to experimental anaemia in the starry flounder, *Platichthys stellatus*. J Exp Biol 82:139–162
Wood SC, Weber RE, Davis BJ (1979b) Effects of air-breathing on acid-base balance in the catfish, *Hypostomus* sp. Comp Biochem Physiol 62A:185–188
Wood SC, Weber RE, Powers DA (1979c) Respiratory properties of blood and hemoglobin solutions from the piranha. Comp Biochem Physiol 62A:163–168
Woodwell GM, Hobble JE, Houghton RA, Mellilo JM, Moore B, Petersen BJ, Shaver GR (1983) Global deforestation: contribution to atmospheric carbon dioxide. Science 222: 1081–1086
Wootton RJ (1979) Energy costs of egg production and environmental determinants of fecundity in teleost fishes. Symp Zool Soc Lond 44:133–159
Wootton RJ (1990) Ecology of teleost fishes. Chapman and Hall, New York, 404 pp
Worthmann HO (1982) Aspekte der biologie zweier Scianidenanten, der pescadas *Plagioscion squamosissimus* (Heckel) und *Plagioscion montei* (Soares), in verschiedenen Gewarssertypen Zentralamazoniens. PhD Thesis, Univ Christi an Albrecht, 176 pp
Wourms JP (1981) Viviparity: the maternal-fetal relationship in fish. Am Zool 21:473–515
Wuntch T, Goldberg E (1970) A comparative physicochemical characterization of lactate dehydrogenase isozymes in brook trout, lake trout, and their hybrid splake trout. J Exp Zool 174:233–252

Yamawaki H, Tsukuda H (1979) Significance of the variation in isozymes of liver lactate dehydrogenase with thermal acclimation in goldfish. I. Thermostability and temperature dependency. Comp Biochem Physiol 62B:89–93

Yano CS (1989) Aspectos hematológicos – série branca de duas espécies de peixes da bacia Amazônica. Internal report, CNPq, Manaus

Zaniboni E (1985) Biologia da reprodução do matrinchã, *Brycon cephalus* (Gunther, 1859) (Teleostei: Characidae). MSc Thesis, PPG INPA/FUA, 134 pp

Zuanon JAS (1990) Aspectos da biologia, ecologia e pesca de grandes bagres (Pisces: Siluriformes, Siluroidei) na área da ilha de Marchantaria – rio Solimões, AM. MSc Thesis, PPG INPA/FUA, Manaus, 186 pp

Systematic Index

Subject Index